Fluid Power
Educational
Series

Electro-pneumatics and Automation

Joji Parambath

Electro-pneumatics and Automation

Copyright © 2023 Joji Parambath

All rights reserved

No part of this book may be reproduced or transmitted in any form or by any means, electronic or mechanical, including photocopying, recording, or by any information storage and retrieval system, without written permission from the publisher.

ISBN: 9798652544386

https://jojibooks.com

First Edition 2020
Revised Edition 2023

Disclaimer of Liability

The contents of this textbook have been checked for accuracy. We cannot guarantee full agreement since deviations cannot be avoided entirely. Only qualified personnel should be allowed to install and work on pneumatic equipment. Qualified persons are those authorized to commission, ground, and tag circuits, equipment, and systems following established safety practices and standards.

Dedicated to

all my amazing students

Table of Contents

Chapter	Description	Page No.
	Preface	vii
1	Introduction to Electro-pneumatic Systems	1
2	Solenoid Valves	4
3	Electrical Control Devices and Control Circuits	8
4	Multiple-actuator Electro-pneumatic Circuits	44
5	Review Questions	62
Appendix 1	Graphic (Symbolic) Representation of Directional Control Valves	63
Appendix 2	Graphic Symbols for Pneumatic Components	65
6	References	72

Preface

In the world of technology, electro-pneumatic systems are becoming increasingly popular. These systems combine the power of pneumatic systems with the control flexibility of electric systems, thus opening up new possibilities for high-performance machines. This hybrid technology incorporates solenoid valves to connect the control and power components. A typical solenoid valve is a converter that generates pneumatic outputs in response to electrical input signals. Elect electro-pneumatic systems utilize control and feedback elements like pushbuttons (PBs), relays, sensors, and timers to achieve the desired control.

This book delves into the workings of solenoid valves and various electrical control components. Numerous single-actuator and multiple-actuator electro-pneumatic circuits are presented to demonstrate various electro-pneumatic applications. The issue of signal conflicts and the various methods of eliminating them in electro-pneumatic circuits are thoroughly explained in this book.

Enjoy reading the book.
Your feedback is most welcome.

Joji Parambath

About the Author….

Joji Parambath is an accomplished expert in Pneumatics, Hydraulics, and PLC with an extensive 25-year background in the field. Over the course of his career, he has trained a multitude of professionals from diverse industries, as well as faculty members and engineering students.

Joji is the primary faculty member at Fluidsys Training Center in Bangalore, India, offering comprehensive training in Pneumatics and Hydraulics. He has authored an impressive 39 books on the subject matter, all designed to convey knowledge on Pneumatics and Hydraulics in a simplistic and easy-to-understand manner.

Joji attributes the creation of his book series to the active engagement and valuable suggestions of his trainees during the training programs. He would like to extend his gratitude towards them.

10th June 2020

Chapter 1 Introduction to Electro-pneumatic Systems

Pneumatic valves can be operated in a few different ways, like manually, mechanically, pneumatically, or electrically. Electric actuation, in particular, relies on a crucial component known as a solenoid to produce the necessary force. Perhaps you are familiar with solenoid valves, a popular example of electro-pneumatic valves.

Electro-pneumatic systems are widely used in industrial machinery and production systems. An electro-pneumatic system combines a pneumatic power transmission system and an electrical control system.

Pneumatic Power Transmission System

A pneumatic power transmission system is shown in Figure 1.1. In this system, compressed air is generated by a compressor when driven by its prime mover, such as an electric motor. The energy in the form of compressed air is then transmitted to actuators, such as cylinders and pneumatic motors, through final control elements, such as directional control valves. The actuators power the work operations required in the system. A pneumatic, relay, or PLC controller controls the final control elements.

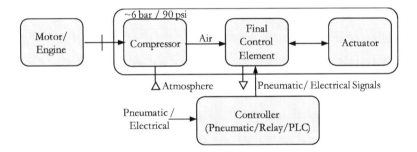

Figure 1.1 | Block diagram of a pneumatic power system

Compressed Air Generation, Storage, and Distribution

The power source in a pneumatic system must be designed to supply a sufficient quantity of compressed air to all the actuators in the system for getting some useful work operations. The primary functions of the power source include the generation and storage of compressed air, regulation of pressure, removal of solid contaminants, moisture, oil particles, and heat from the compressed air, and distribution of the compressed air to the system. A schematic diagram and symbolic representations of the power supply are given in Figure 1.2.

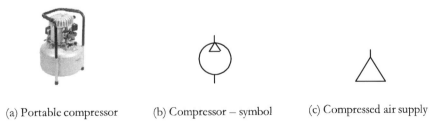

Figure 1.2 | A schematic and symbols of the pneumatic power supply

Pneumatic Actuators

Pneumatic actuators are output devices that convert the energy contained in compressed air to produce linear or rotary motion or apply a force. They are devices for providing power and motion to automated systems, machines, and processes. Pneumatic actuators can be categorized into two basic types: (1) Linear actuators and (2) Rotary actuators.

Linear actuators are devices that convert pneumatic energy into straight-line mechanical energy. They come in two main types: single-acting cylinders and double-acting cylinders. On the other hand, rotary actuators convert pneumatic energy into rotary mechanical energy. Examples of rotary actuators include semi-rotary actuators and pneumatic motors. Semi-rotary actuators are designed for reciprocating rotary motion up to 360°, while air motors are engineered for continuous rotation.

Single-acting Cylinder

Figure 1.3 shows the cross-sectional view of a single-acting pneumatic cylinder. It consists of a barrel, a piston-and-rod assembly, a spring, end caps, seals, and a port. The piston-and-rod assembly is a tight fit inside the barrel and is biased by the spring. A fluid chamber is formed in the cylinder with the barrel, the piston, and the cap-end endplate.

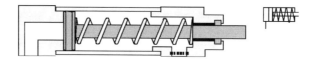

Figure 1.3 | Single-acting cylinder

When pressure is exerted through the port, the piston-and-rod assembly is pushed in one direction to create the working stroke. However, the assembly moves in the opposite direction due to an external force, spring force, or gravity. This type of cylinder can only generate work in one direction of its movement. Hence it is known as a single-acting cylinder.

Double-acting Cylinder

Figure 1.4 gives the cross-sectional view of a double-acting pneumatic cylinder. It mainly consists of a barrel, a piston-and-rod assembly, end caps, seals, and two ports. The double-acting cylinder has fluid ports on both ends, namely the piston-side port and the piston-rod-side port.

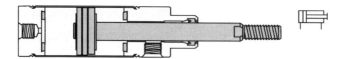

Figure 1.4 | Double-acting cylinder

By exerting pressure on the piston-side port, the cylinder extends as long as the pressure is relieved from the piston-rod side. Similarly, applying pressure on the piston-rod-side port retracts the cylinder, provided that the pressure from the piston side is released. This process allows the cylinder to transform pressure and flow into force and motion. A double-acting cylinder can execute work operations in both directions of its motion, which is why it is called a 'double-acting cylinder'.

Semi-rotary Actuators

Semi-rotary actuators are designed to execute reciprocating rotary motion up to 360°. They can be built with either a rotating vane or a rack-and-pinion design. Figure 1.5 illustrates a rack-and-pinion type rotary actuator, which includes a double-acting piston linked to the output shaft through a rack-and-pinion mechanism.

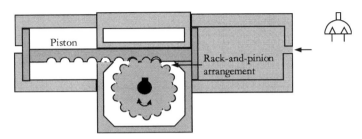

Figure 1.5 | Semi-rotary actuator

Air Motors

Air motors convert the potential energy of the compressed air into rotary mechanical energy. They are designed to provide continuous rotation. A vane type pneumatic motor is shown in Figure 1.6.

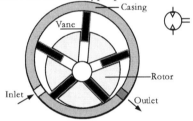

Figure 1.6 | Air motor

Electro-Pneumatic Systems

In electro-pneumatics, an electrical control system controls a pneumatic power transmission system. A final control element, such as a solenoid valve, acts as the interfacing component between the pneumatic power system and the electrical control system. The control system mainly consists of an electromagnetic relay controller and many input devices such as pushbuttons and sensors. In the relay controller, many relays are interconnected, as per the governing logic, to control the final control elements (solenoid valves) and achieve the desired control function. In complex applications, electrical controls are almost exclusively employed for the optimum utilization of both types of energy media for cost-effective and efficient production systems.

This book comprehensively explains solenoid valves, pushbuttons (PBs), relays, sensors, timers, pressure switches, and counters, all crucial electrical control components. Additionally, the book covers various sensors and showcases the development of many electro-pneumatic circuits that utilize these sensors to automate electro-pneumatic systems. This book aims to help readers understand relay circuits and their automation with practical examples. The relay circuits presented in this book progressively increase in complexity, and it is recommended that readers work through them in the order presented.

Chapter 2 Solenoid Valves

Solenoid - Fundamentals

The magnetic effect of electric current can be used for implementing various technical functions in industrial pneumatic systems. When an electric current passes through a straight conductor, magnetic lines of force are produced around the conductor, as shown in Figure 2.1(a). The direction of these lines of force depends on the direction of the current flow. However, this magnetic force is distributed over the length of the conductor. Hence, it cannot be utilized for realizing any purposeful control function.

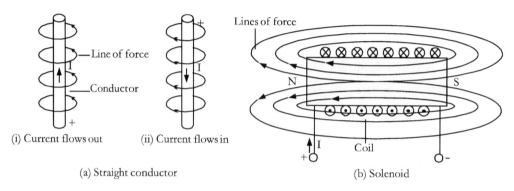

Figure 2.1 | A graphic representation showing the magnetic effect of electric current

Next, the conductor is wound as a long cylindrical coil to concentrate the magnetic lines of force. The arrangement of the coil, as shown in Figure 2.1(b), is usually known as a solenoid. A magnetic field surrounds the coil when the current passes through it. The strength of the magnetic field depends on the following factors: (1) the number of turns of the coil, (2) the magnitude of the current, and (3) the nature of the core material. A soft iron core is introduced into the solenoid to strengthen the magnetic force further.

The off-center soft iron core of the solenoid is also useful in developing the actuating force in a solenoid valve. When an electric current passes through the off-center core, it is pulled toward the center of the coil, generating a linear force that can be used to actuate the valve.

AC solenoids Vs DC Solenoids

A solenoid consists of a coil and a movable iron core. Solenoid coils can generally be designed to operate on AC or DC power. Many differences can be recognized between the AC and DC solenoids. A few of them are outlined below.

The core of a DC solenoid consists of a soft iron providing optimum conductance for the magnetic field. However, the operation of solenoids with AC currents introduces hysteresis and eddy current losses in the core. The armature of the AC solenoid consists of laminated metal sheets to reduce iron losses. However, there is still a substantial temperature rise when the AC solenoid operates. The AC solenoid also develops a high initial inrush current. If the armature in the AC solenoid is not allowed to shift entirely to its end position, the current drawn by it tends to remain high. The result is that the coil overheats and burns out. In contrast, the DC solenoid does not experience much inrush current. Therefore, the armature of the DC solenoid can remain partially shifted to a particular position indefinitely without an increase in the current drawn by it.

Other Miscellaneous Solenoid Characteristics

DC solenoids are characterized by quiet switching action, requiring low turn-on and holding power. AC solenoids are characterized by short switching times and the development of a large pulling force. However, DC solenoids are subjected to sizeable induced contact wear and over-voltages during supply cut-off. An AC solenoid is designed with a lower resistance than that of a DC solenoid because an AC solenoid's inductance will also limit the current.

Solenoid Valves

A solenoid valve interfaces the pneumatic power and electrical control parts of an electro-pneumatic system. The valve can be of the discrete type or infinitely variable type. In discrete valves, the spool occupies two or three discrete positions when actuated. In an infinitely variable valve, the output can be infinitely varied in relation to the applied input signal.

A discrete electro-pneumatic solenoid valve consists of pneumatic power and electrical control parts. The pneumatic section includes a valve body and a spool for the direction control, and the electrical part consists of a solenoid and a core (plunger) for linking and controlling the pneumatic section. The core is usually positioned away from the center of the coil by a biasing spring. When the coil is energized, the resultant magnetic field pulls the core towards the center of the coil. This movement of the core is used to actuate the solenoid valve. Thus, the incoming electrical signal at the input of the valve can be converted to the corresponding fluid flow at the output of the valve.

Discrete electro-pneumatic solenoid valves can be classified as: (1) direct-acting type and (2) pilot-operated type. In the direct-acting type solenoid valve, the solenoid and its plunger open an orifice in the valve. The pilot-operated solenoid valve consists of an internal pilot valve and the main valve. In this type of valve, a pilot electrical signal controls the opening of the pilot valve, which in turn provides the actuating force for the main valve. Further, electro-pneumatic solenoid valves can be classified according to the type of 'port/position' configuration, such as 3/2-way, 5/2-way, and 5/3-way directional control valves.

3/2-way Single-solenoid Valve, Spring Return

Figure 2.2 shows the cross-sectional views of a 3/2-way single-solenoid electro-pneumatic valve in its normal and actuated positions. The valve mainly consists of an electrical part controlling a pneumatic power part. Further, the pneumatic part is a body with a spool inside, a reset spring, and necessary ports (1, 2, and 3), and the electrical part consists of a solenoid (Y) with an armature.

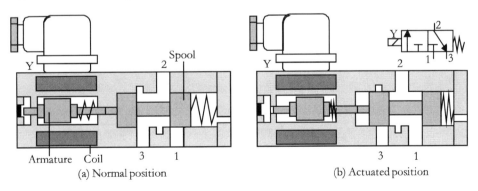

Figure 2.2 | Cross-sectional views of a 3/2-way single-solenoid pneumatic valve

When the solenoid valve is in its normal position [as shown in Figure 2.2(a)], pressure port 1 is blocked, and working port 2 is internally connected to exhaust port 3. Upon applying the rated voltage to solenoid coil Y, the solenoid's armature moves towards the center of the coil, causing the spool to shift away from the valve seat. This movement of the armature puts the valve in its actuated position.

In the actuated position [Figure 2.2(b)], pressure port 1 is internally connected to working port 2, and exhaust port 3 is blocked. Once the coil's supply is cut off, the valve reverts to its normal position.

The manual override facility is a standard feature on a solenoid valve. This feature can be used to manually operate the valve by turning an eccentric screw provided. Further, a light-emitting diode (LED) can be incorporated into the housing of the solenoid valve for the visual indication of the ON/OFF state of the coil.

The 3/2-DC solenoid valve can be the final control element for controlling a single-acting cylinder.

5/2-way Single-solenoid Valve, Spring Return

Figure 2.3 shows the cross-sectional views of a 5/2-way single-solenoid valve in its normal and actuated positions. The valve comprises two parts - an electrical component that controls a pneumatic power component. The pneumatic part includes a body with a spool, a reset spring, and ports numbered 1 through 5. The electrical part consists of a solenoid (Y) with an armature.

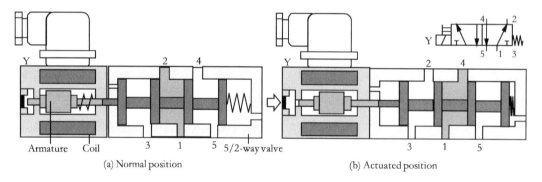

Figure 2.3 | Cross-sectional views of a 5/2-way single solenoid valve

In the normal position of the solenoid valve [Figure 2.3(a)], pressure port 1 is connected to working port 2, and working port 4 is connected to exhaust port 5. The valve is actuated when the rated voltage is applied to coil Y.

In the actuated position of the valve [Figure 2.3(b)], pressure port 1 is connected to working port 4, and working port 2 is connected to exhaust port 3. When the supply to the coil is cut off, the valve returns to its normal position.

The 5/2-DC solenoid valve can be the final control element for controlling a double-acting cylinder.

Note: The concepts of the symbolic representations of electro-pneumatic valves are given in Appendix 1. Symbols of pneumatic components are given in Appendix 2.

5/2-way Double-solenoid Valve

Figure 2.4 shows the cross-sectional views of a 5/2-way double-solenoid electro-pneumatic valve. The valve mainly consists of an electrical part controlling a pneumatic power part. Further, the pneumatic part is a body with a spool inside, a reset spring, and necessary ports (1, 2, 3, 4, and 5), and the electrical part consists of two solenoid coils (Y1 and Y2) on either side of the valve. This type of valve design is noted for the absence of a reset spring.

When the rated voltage is applied to coil Y1 momentary or continuously, the valve is actuated to a particular switching position. In this switching position (not shown), pressure port 1 is connected to working port 4, and working port 2 is connected to exhaust port 3. The valve maintains this position until a signal is applied to coil Y2. When the rated voltage is applied to coil Y2 momentary or continuously, the valve is actuated to the alternative switching position. In this switching position (Figure 2.4), pressure port 1 is connected to working port 2, and working port 4 is connected to exhaust port 5. The valve maintains this position until a signal is applied to the other coil Y1.

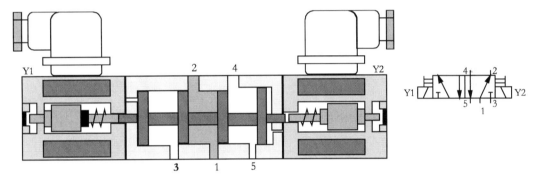

Figure 2.4 | A cross-sectional view of a 5/2-way double solenoid valve in its normal position

The valve remains in a particular position due to applying a pulse or continuous signal to the coil at one end as long as no opposing signal is present at the other end. Hence, this valve shows memory characteristics. This valve can be the final control element for controlling a double-acting cylinder.

Further, it can be seen that when control voltages are applied to both coils simultaneously, the valve cannot switch its current position. This problem is known as 'signal conflict'.

5/3-way Solenoid Valves

5/3-way valves can be designed with many types of center positions. The most common types of center positions are: (1) All closed center position, (2) Exhaust center position, and (3) Pressure center position, as shown in the self-explanatory Figure 2.5.

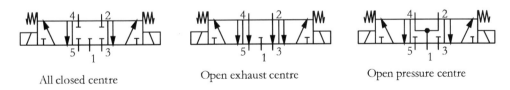

All closed centre　　　Open exhaust centre　　　Open pressure centre

Figure 2.5 | 5/3-way electro-pneumatic pneumatic valves with different center positions

Chapter 3 Electrical Control Devices and Control Circuits

Often, a technician is overwhelmed by the complexity and size of control systems for industrial applications like automatic assembly lines. However, this complexity is fine for those who understand the fundamentals of control and the principle of operation of essential control components. Even a complex control system comprises a series of simple control circuits involving basic control components. Some basic control components are: pushbuttons, limit switches, float switches, pressure switches, flow switches, thermostats, relays, proximity sensors, timers, and counters.

The functional aspects of many control components and some typical control circuits are presented in the following sections. A list of important graphic symbols used for electrical components is given in Appendix 1.

Switch

An electrical switch consists mainly of control contacts for making or breaking an electrical circuit. In control applications, switches are integrated as control contacts in various pilot devices, such as pushbuttons, limit switches, pressure switches, timers, and counters. The purpose of control contacts is to present electrical signals from various points in the control system to the area of signal processing.

Pushbuttons

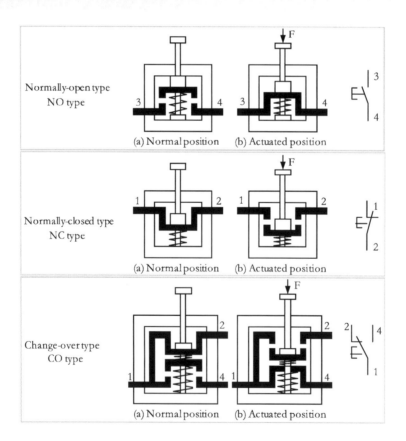

Figure 3.1 | Types of pushbuttons and their symbols

A pushbutton is a switch to close or open an electric control circuit. This device consists of fixed and movable contacts and a restraining spring. Pressing the pushbutton against the restraining spring operates its contacts.

Pushbuttons are of two types:

- Momentary-contact type
- Maintained-contact type (or detent type)

In the momentary-contact type, the contacts are operated only when the pushbutton is pressed, and the contacts return to their normal position when the pushbutton is released.

In the maintained-contact type, the contacts are operated when the pushbutton is pressed. The contacts remain in place even after the actuating force is removed. The contacts return to their original position when the pushbutton is pressed again.

Contact Types
The contacts of the pushbuttons, distinguished according to their functions, are as follows:

- Normally Open (NO) type
- Normally Closed (NC) type
- Change-Over (CO) type.

The cross-sections of various types of pushbuttons in the normal and actuated positions and their symbols are given in Figure 3.1.

In the NO type, the contacts are open in the normal position, inhibiting the energy flow through them. In the actuated position, the contacts are closed, permitting the energy flow through them.

In the NC type, the contacts are closed in the normal position, permitting the energy flow through them. In the actuated position, the contacts are open, inhibiting the energy flow through them.

Changeover contact is a combination of NO and NC contacts.

Terminal Markings of Contacts
Contacts are used in many pilot devices, such as pushbuttons, relays, timers, and counters. As a standard practice, contact terminals are designated with a set of numbers based on the contact function to identify the terminals. The numbering system for the contact terminals is given in Table 3.1

Table 3.1 | Terminal markings of electric contacts

Type of pilot device	Terminal numbers of	
	NC contact	NO contact
Ordinary devices (PBs, relays)	1 and 2	3 and 4
Special devices (Timers, counters)	5 and 6	7 and 8

Pushbutton Station

A pushbutton station consists of many contacts (NO, NC, or CO) pairs/sets with a common actuation, as shown in Figure 3.2(a). This device is compact and less expensive but usually has limited current carrying capacity.

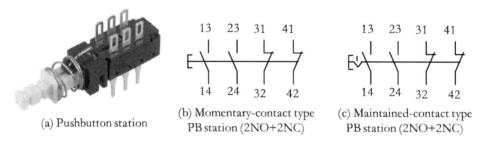

(a) Pushbutton station

(b) Momentary-contact type PB station (2NO+2NC)

(c) Maintained-contact type PB station (2NO+2NC)

Figure 3.2 | Symbolic representations of Pushbutton stations.

Figure 3.2(b) shows the symbol of a momentary-contact type pushbutton station with a contact configuration of 2 NO + 2 NC.

Figure 3.2(c) shows the symbol of a maintained-contact type pushbutton station.

A pair of consecutive two-digit numbers are used to designate the terminals of a contact in the pushbutton station. In the two-digit number, the unit figures indicate the function of the contact (that is, whether it is a NO or NC type). The digits at the ten's place merely represent a serial ordering of all contact pairs in the pushbutton station for identifying each contact pair uniquely.

Let's take an example of unit figures 3-4 in the 13-14 designation for the terminals of the first contact. These figures imply that it is a NO contact, and the digit 1 in the ten's place is used for the serial numbering to identify the contact uniquely. In the same way, the designations of other terminals can be interpreted.

In pushbutton stations, the actuating force increases as the number of contacts increases. For this reason, the number of contacts is usually limited in pushbutton stations.

Industrial Control Voltages

In earlier days, the control voltages used in industries were 230 V AC, 110 V AC, etc. However, the tendency was to reduce the control voltage to a lower level from the operator's safety point of view.

Currently, 24 V DC is the standard industrial control voltage in most countries. A DC power pack can convert 230 V AC to 24 V DC. The possible voltage levels for electrical control components used in industrial systems are: 12 V DC, 24 V DC, 24 V 50/60 Hz, 48 V 50/60 Hz, 110/120 V 50/60 Hz, and 220/230 V 50/60 Hz.

This textbook shows all the electro-pneumatic control circuits with 24 V DC control voltage.

Example 3.1 | Direct control of a single-acting cylinder

A small-volume single-acting pneumatic cylinder should extend when a pushbutton (PB) is pressed. The cylinder should retract when the pushbutton is released. A 3/2-way single solenoid valve, rated for a coil voltage of 24 V DC, controls the cylinder. Develop a pneumatic power circuit and an electrical control circuit to implement the control task.

Solution

An electro-pneumatic circuit diagram is conventionally drawn with two distinctive parts. First, the pneumatic power circuit is drawn, and then the electrical control circuit is drawn just below the pneumatic circuit. The interface between the pneumatic and electrical elements is the solenoid coil Y that appears on the pneumatic and electrical circuits with a common designation.

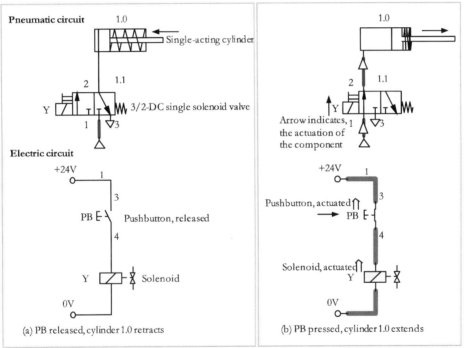

Figure 3.3 | Two circuit positions for the direct control of a single-acting cylinder (Example 3.1)

The desired control task by the single-acting cylinder designed for spring-assisted retraction is illustrated with two circuit positions in Figure 3.3. The solenoid valve is shown with a manual override.

The actuation of the pushbutton (PB) generates current flow through solenoid coil Y, which in turn causes the actuation of valve 1.1, as shown in Figure 3.3(b). The compressed air flows from port 1 to port 2 of the valve, and the cylinder extends against the spring force.

When the pushbutton is released, the electrical circuit is interrupted. The solenoid coil is de-energized, and the valve returns to its original position, as shown in Figure 3.3(a). The compressed air in the cylinder then exhausts through port 3 of the valve, and the single-acting cylinder retracts with the help of the spring.

Example 3.2 | Direct control of a double-acting cylinder

A small-volume double-acting pneumatic cylinder should extend when a pushbutton (PB) is pressed and should retract when the pushbutton is released. A 5/2-way single solenoid valve controls the extension and retraction of the cylinder. Develop a pneumatic power circuit and an electrical control circuit to implement the control task.

Solution

The multiple positions of the circuit for the desired control task by the double-acting cylinder are illustrated in Figure 3.4. The pushbutton (PB) actuation generates a current flow through solenoid coil Y and actuates valve 1.1, as shown in Figure 3.4(b). The compressed air then flows from port 1 to port 4 of the valve and reaches the piston side of the cylinder. The compressed air from the other side of the cylinder exhausts through port 3 of the valve. As a result, the cylinder extends to its forward end position.

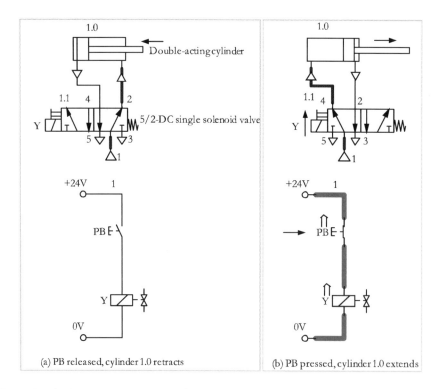

Figure 3.4 | Two circuit positions for the direct control of a double-acting cylinder (Example 3.2)

When the pushbutton is released, the electrical circuit is interrupted. The solenoid coil is de-energized, and the valve returns to its original position, as shown in Figure 3.4(a). The supply air flows from port 1 to port 2 of the valve and reaches the piston-rod side of the cylinder. The compressed air from the piston side of the cylinder exhausts through port 5 of the valve. As a result, the cylinder retracts to its rear-end position.

Looking at the examples in 3.1 and 3.2, it is evident that controlling single-acting and double-acting cylinders using pushbuttons is straightforward. However, let's proceed to more complex controls.

Electromagnetic Relay

A relay can be considered as an electro-magnetically actuated switch that operates under the control of an additional electrical circuit. It is a simple electrical device used mainly for signal processing. This switch is designed to withstand heavy power surges and harsh environmental conditions.

The cross-section of a relay and its symbol are shown in Figure 3.5. The relay consists of a coil and a few stationary and movable contacts. It also consists of a stationary core and a movable core to conduct the magnetic field. It is to be noted that the fixed core is placed inside the coil to strengthen the magnetic field. The movable contacts are coupled to the movable core.

Therefore, when the coil is energized, the movable core is pulled to the stationary core, thus simultaneously operating all its coupled contacts. This movement either makes or breaks the connection with the respective fixed contact. Thus, a relay is a combination of a coil and contacts. Further, the contacts are operated when the coil is energized.

A large number of control contacts can be incorporated into relays. Remember, relays are usually designated as K1, K2, K3, etc.

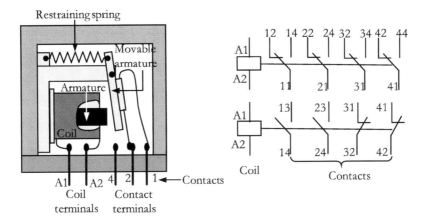

Figure 3.5 | Cross-sectional view of a relay

Relays are generally used to accept information from pilot devices like sensors and convert this information into a proper power level. Because a relay can govern an output circuit with higher power than an input circuit, it can broadly be regarded as an electrical amplifier.

Relays can adapt themselves to many control applications, including automation, because of their low cost, easy adaptability, and high operating speed. They can perform many control functions required by the basic automatic controls.

A relay also possesses the interlocking capability, regarded as an important safety feature in control circuits. This feature must be incorporated in situations where the simultaneous switching of individual coils will result in short circuits or other undesirable operations. Relays with interlocking capability can be used to prevent the simultaneous switching-on of the coils accidentally.

Example 3.3 | Relay control

A large volume double-acting cylinder is to be controlled by using a 5/2-way single solenoid valve. The cylinder should extend when the PB is pressed and should retract when the PB is released. Develop an electro-pneumatic control circuit to implement this control task.

Solution

The operation of the circuit required for the given control task is given in Figure 3.6. The actuation of pushbutton PB energizes relay coil K, and consequently, all of its contacts are operated. Solenoid coil Y is energized through the NO contact of the relay in branch 2, causing the actuation of valve 1.1, as illustrated in Figure 3.6(b). The cylinder then extends to its final forward-end position.

When pushbutton PB is released, the electrical circuit in branch 1 is interrupted. The relay and solenoid coils are de-energized, and the valve returns to its original position, as illustrated in Figure 3.6(a). The cylinder then retracts to its rear-end position.

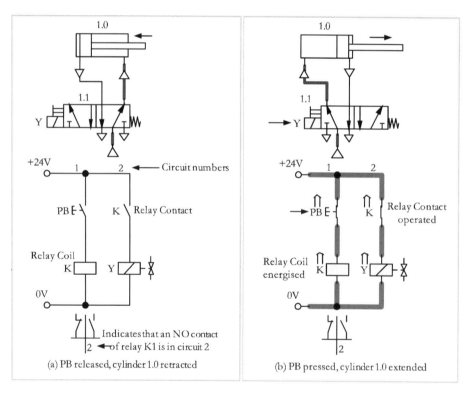

Figure 3.6 | Two circuit positions for the indirect control of a double-acting cylinder using a relay (Example 3.3)

It is usual to show the consolidated information of the positions of the control contacts of relay K in the circuit diagram just below the symbol of the relay coil. For example, the use of an NO contact of relay K in branch 2 is indicated below the coil, as shown in Figure 3.6.

Logic Controls

Logic circuits are designed to perform decision-making output functions based on many input signals from pushbuttons or sensors representing certain conditions of the associated machines or systems. Two essential logic functions are: 'AND' and 'OR'. Signal levels in logic devices are characterized by two states: logic '1' and logic '0'. Usually, the logic '1' represents an ON state, and the logic '0' represents an OFF state. The block diagrams and truth tables for the OR function are given in Figure 3.7(a), and those for the AND function are in Figure 3.7(b).

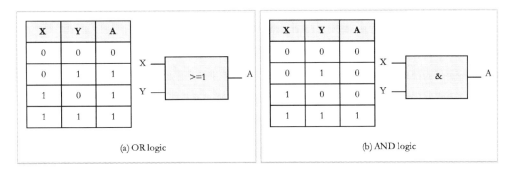

Figure 3.7 | Block diagrams and truth tables for 'OR' and 'AND' logic functions

The OR logic function has an output when one or more inputs are present. An example of the OR logic function is a lamp controlled from one or more positions. A parallel connection of input devices makes an OR function. Its operation is illustrated in the self-explanatory Figure 3.8(a) for controlling a lamp using two pushbuttons connected in parallel.

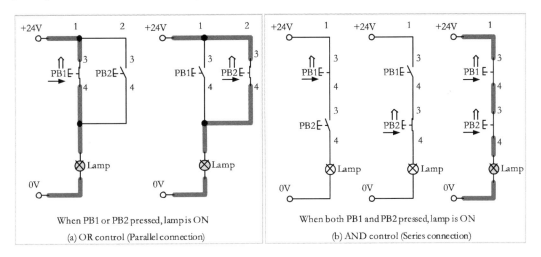

Figure 3.8 | Logic controls of a lamp using two pushbuttons

In the AND logic function, there is an output only when all the inputs are present. An example of the AND logic function is the requirement of a cylinder that extends only when its safety guard is placed in position and a 'start' signal is given. A series connection of input devices like pushbuttons and sensors gives an AND function. Its operation is illustrated in Figure 3.8(b) for controlling a lamp using two pushbuttons connected in series. Remember, any complex logic function can also be set up easily by a series-parallel connection of input devices.

Example 3.4 | Control of a double-acting cylinder from two locations

A large-volume double-acting pneumatic cylinder should be controlled from two pushbuttons (PB1 and PB2) installed at different locations. The cylinder should extend when either PB1 or PB2 is pressed. The cylinder should retract and remain in the home position when PB1 and PB2 are released. Develop an electro-pneumatic circuit to implement the control task.

Solution

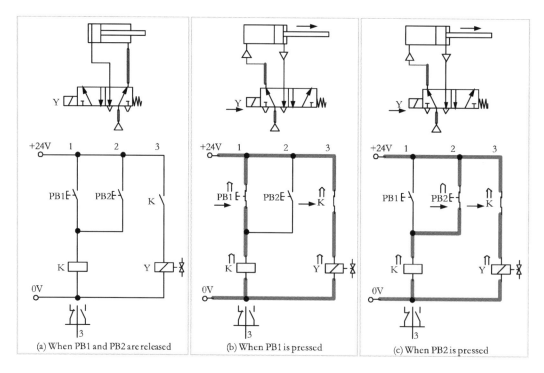

Figure 3.9 | Logic control of a double-acting cylinder from two locations (Example 3.4)

The electro-pneumatic circuit in three critical positions is given in Figure 3.9.

As shown in the pneumatic part of the circuit, the double-acting cylinder is controlled by a 5/2-way single solenoid spring-return valve. The cylinder extends when the solenoid gets energized.

In the electrical control part, as shown in Figure 3.9(a), the pushbuttons PB1 and PB2 are connected in parallel to realize the OR logic function.

As shown in Figure 3.9(b) and (c), the solenoid Y gets energized through the relay when PB1 or PB2 is pressed.

The result remains the same when both pushbuttons PB1 and PB2 are pressed.

Example 3.5 | Two-hand safety

A large-volume double-acting pneumatic cylinder is used in a machine to punch workpieces. A design objective is to have two-hand safety by using two pushbuttons, PB1 and PB2, installed at a distance to engage the operator's hands on the two pushbuttons. That means the operator cannot operate both pushbuttons simultaneously using one hand. The cylinder should extend and carry out the punching operation when both pushbuttons are pressed. The cylinder should retract when anyone or both pushbuttons are released. Develop an electro-pneumatic circuit to implement the control task.

Solution

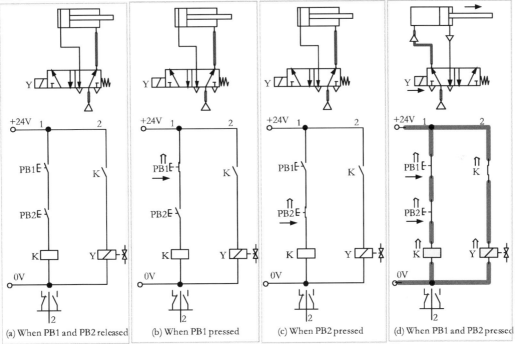

Figure 3.10 | Logic control of the double-acting cylinder for two-hand safety (Example 3.5)

The electro-pneumatic circuit in four critical positions is given in Figure 3.10. In the pneumatic part of the circuit, the double-acting cylinder is shown as controlled by a 5/2-way single solenoid spring-return valve. The cylinder extends when the solenoid Y gets energized. In the electrical control part, pushbuttons PB1 and PB2 are connected in series to realize the AND logic function. Figure 3.10(d) shows that the solenoid gets energized through the relay when both PB1 and PB2 are pressed. The cylinder then extends. For all other conditions, as shown in Figure 3.10(a), (b), and (c), the solenoid remains in the de-energized state.

Memory Function

A memory function 'remembers' the state of the last output even after the input signal (ON) responsible for this output has been removed. An input signal (ON) must be given to set the memory function. To 'reset' the memory function, another input signal (OFF) needs to be given. The normal control of a lamp using a mechanically latched toggle switch in our home is a familiar example of memory control. When we press the switch, it turns to one side and switches the lamp to the ON state. The lamp remains

ON until we press the switch again to turn it the other way. The lamp is switched to the OFF state, and it remains in that state till another ON input signal is given. An electrical latching circuit, as presented in Example 3.6, is another example of the memory function. A memory function can also be realized using a double solenoid valve.

Example 3.6 | Latching circuit

A double-acting cylinder is to be controlled using a 5/2-way single solenoid valve. When pushbutton PB1 is pressed, the cylinder should extend and remain in the extended position even when PB1 is released. The cylinder should retract to the home position when pushbutton PB2 is pressed. The cylinder is to remain in the home position even when PB2 is released. Develop an electro-pneumatic control circuit with an electrical latching circuit.

Solution

Two critical positions of the pneumatic and electrical parts of the electro-pneumatic circuit, when coil Y is energized and de-energized, are given in Figure 3.11. In the pneumatic part, a double-acting cylinder is controlled by a 5/2-way single solenoid valve. The valve does not exhibit memory characteristics; it is reset with a spring. The memory function can be implemented through the electrical latching circuit. The latching circuit, as shown in Figure 3.11(a), consists of a relay (K) and an NO type 'Start' pushbutton (PB1), and an NC type 'Stop' pushbutton (PB2).

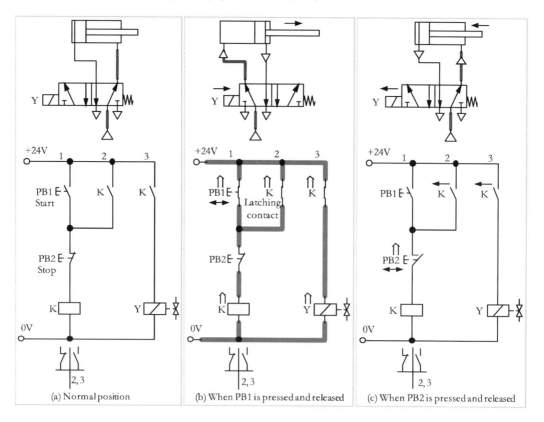

Figure 3.11 | Three critical positions of the electro-pneumatic circuit (Example 3.6)

When the 'Start' pushbutton is momentarily or continuously pressed, relay coil K1 in branch 1 is energized, operating all of its contacts. The NO contact of relay K1, used in branch 2, latches 'Start' pushbutton PB1 by providing a parallel path for the current flow so that relay coil K1 remains energized even when pushbutton PB1 is released. This latched position of the circuit is shown in Figure 3.11(b). The contact of relay K1, used in branch 3, switches solenoid valve Y. The cylinder extends and remains in that position until the 'Stop' pushbutton PB2 is pressed.

When the 'Stop' pushbutton PB2 is momentarily or continuously pressed, relay coil K1 is de-energized as the relay coil circuit is interrupted. All the control contacts of relay K1 return to the normal position. The opening of the relay contacts unlatches the circuit and de-energizes the solenoid. The cylinder retracts and remains in that position until the 'Start' pushbutton PB1 is pressed again. Figure 3.11(c) shows the circuit's un-latched position.

Dominant ON and Dominant OFF Latching Circuits

Two variants of the latching circuits exist according to the position of the Stop pushbutton PB2 in the circuit. They are: (1) Dominant OFF circuit and (2) Dominant ON circuit. The dominant OFF circuit, shown in Figure 3.12(a), remains in the OFF state when both 'Start' pushbutton PB1 and 'Stop' pushbutton PB2 are pressed simultaneously. The dominant ON circuit, shown in Figure 3.12(b), remains ON when both 'Start' pushbutton PB1 and 'Stop' pushbutton PB2 are pressed simultaneously. It may be noted that when the pushbuttons are pressed individually, the circuits operate in the normal way.

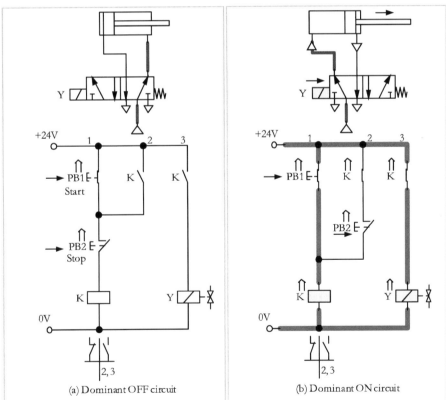

Figure 3.12 | Two types of electrical latching circuits

Example 3.7 | Memory control by using a double-solenoid valve

A double-acting cylinder is to be controlled using a 5/2-way double-solenoid valve. When pushbutton PB1 is pressed, the cylinder should extend and remain in this position even when pushbutton PB1 is released. When pushbutton PB2 is pressed, the cylinder should retract to the home position. The cylinder has to remain in the retracted position even when the pushbutton PB2 is released. Develop an electro-pneumatic control circuit for this operation.

Solution

In Example 3.6, the memory was implemented in the electrical part using the latching circuit. A memory function can also be implemented in the pneumatic part using a double-solenoid valve, as shown in Figure 3.13(a).

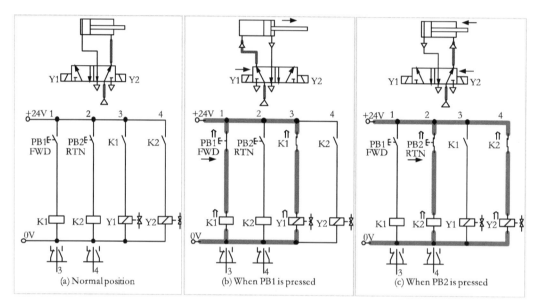

Figure 3.13 | Three critical positions of the electro-pneumatic circuit (Example 3.7)

When pushbutton PB1 is pressed [Figure 3.13(b)], coil Y1 is energized through relay K1 and valve 1.1 switches over. The cylinder extends and remains in the forward-end position until a signal is applied in the opposite direction.

When pushbutton PB2 is pressed [Figure 3.13(c)], coil Y2 is energized through relay K2 and valve 1.1 switches over. The cylinder retracts to its home position until a signal is applied to coil Y1 again.

In short, a signal to any of the coils switches the valve to the corresponding position. The valve remains in that switching position until a signal is applied to the other coil.

However, it should be remembered that double-solenoid valves are susceptible to the problem of signal conflicts.

Sensors

We can only think of modern industrial production systems with sensors. Various sensors are devised to meet the varied demands of industrial production systems. Sensors react to changes in conditions around them by generating signals in a form that a control system can process.

An electrical (or pneumatic) output signal is generated in a sensor in response to a disturbance caused to some physical medium (like magnetic, electric, optic, or acoustic) by an approaching object or by the influence of an internal change.

The cause of disturbance may be the variation in pressure, flow, temperature, force, liquid level, or a particular position of a cylinder piston or workpiece.

Classification of Sensors

A sensor senses the presence of an object by the actual physical contact with the object or by the object's movement in close proximity. Accordingly, sensors are classified as:
- Contact type sensors (e.g., limit switch)
- Contactless type sensors (e.g., proximity sensor).

Sensors are also classified, depending on whether they deliver discrete signals or analog signals, as:
- Discrete sensors
- Analog sensors

A discrete sensor converts a physical quantity into a binary (digital) signal. Discrete sensors include limit switches, proximity sensors, and pressure switches.

An analog sensor converts a physical quantity, such as flow, force, torque, pressure, or length, into an analog electrical signal, such as voltage or current proportional to the sensed input parameter. Examples of analog sensors are temperature, level, and pressure sensors.

Drawbacks of Sensors

Wear and 'contact bounce' are significant problems in contact-type sensors.

Another problem with sensors with switching contacts when used in circuits with inductive loads is the development of high voltage peaks at the moment of circuit cut-off due to the breaking action of the switching contacts.

Therefore, protective circuits are essential in sensors with switching contacts. An RC element, diode, or a varistor of appropriate rating may be integrated into the sensor for protection.

Applications of Sensors

Sensors find applications in sequence control of a production system, monitoring of safety devices in production systems, detection of faults, open-loop control technology, and closed-loop control technology.

Some important sensor types are explained below:

Limit Switch

Limit switches perform the same way as pushbuttons. However, the main difference between the two types is that pushbuttons are actuated manually, whereas limit switches are actuated mechanically. A limit switch indicates a particular final position of a machine part or cylinder piston. It comprises a set of contacts and a roller lever linked to these contacts, as shown in Figure 3.14. Electrical contact is established or interrupted using an external force acting on the roller lever.

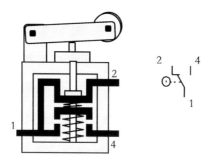

Figure 3.14 | Limit switch

Limit switches can be distinguished according to their actuation methods, i.e., lever-actuated or spring-loaded contacts. In a lever-type limit switch, the contacts are operated slowly. Hence, this type of limit switch is suitable for slow approach speeds. With a spring-loaded type limit switch, the approach speed is immaterial because the switch changes its state rapidly.

Further, the contacts in limit switches or other control switches tend to bounce when operated. The contact bounce is not a serious problem when a fan or a lamp is turned on. However, the effects of contact bounce must be considered when a switching contact is used as input to a digital counter, a personal computer, or a microprocessor-based piece of equipment.

Reed Switch

A reed switch is also known as a magnetically actuated proximity switch. Its cross-sectional view and symbol are shown in Figure 3.15(a).

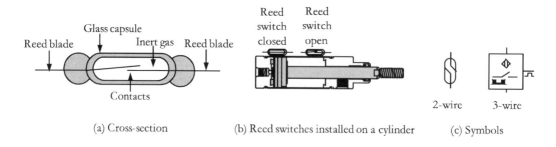

(a) Cross-section   (b) Reed switches installed on a cylinder   (c) Symbols

Figure 3.15 | Reed switch

A reed switch consists of two springy metal strips acting as switching contacts, hermetically sealed in a glass tube filled with an inert gas to prevent corrosion. This unit is further encapsulated in epoxy resin to prevent any mechanical damage. The switching contacts (soft magnetic metallic reeds) are usually made of an alloy of iron and nickel. A reed switch is provided with an LED indicator to show its switching status. It is designed for mounting on a cylinder [Figure 3.15(b)] and reacts to the magnetic fields of the permanent magnets invariably provided on the cylinder piston. An output signal is produced at the reed switch when the piston is close enough for the magnetic field to actuate the contacts of the switch.

The primary reed switch consists of only two wires – one for the connection to the positive terminal of the electric supply and the other for taking out the signal output. The three-wire reed switch consists of three wires – one for the connection to the positive terminal of the electric supply, the second for taking out the signal output, and the third for the negative terminal of the electric supply (necessary for the LED indicator). Regarding advantages, reed switches are compact, reliable, and wear-free. However, the closing of contacts in reed switches is not bounce-free. The symbols of different types of reed switches are given in Figure 3.15(c).

Proximity Sensors

Discrete output proximity sensors are the most important for industrial applications. They gain importance in such applications where the recording or counting of moving objects or workpieces on machines or conveyors is necessary. In this type of application, a limit switch cannot be used because the actuating force of the workpiece may not be adequate to trigger the switch. Similarly, in situations where the sensing point of a workpiece is not relevant to the position of a cylinder piston, a reed switch is unsuitable for use. Hence, proximity sensors with contactless sensing are ideally suited to these applications. These sensors are also available in universal voltage designs; hence they can be connected to either DC or AC voltage.

There are three types of proximity sensors: (i) Inductive type, (ii) Capacitive type, and (iii) Optical type.

Inductive Proximity Sensor

An inductive proximity sensor consists of the following blocks: (1) an oscillator (LC resonant circuit), (2) a switching circuit, (3) an amplifier, and (4) an output stage. An inductive proximity sensor with these blocks is shown in Figure 3.16.

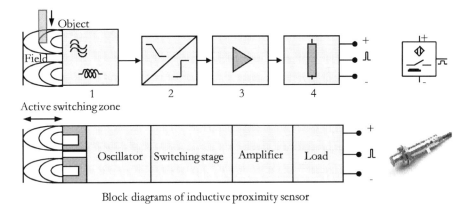

Block diagrams of inductive proximity sensor

Figure 3.16 | An inductive proximity sensor

The sensing surface is designed with a half-open shell made of ferrite material. This shell creates a magnetic field across the active surface of the sensor through a small distance called an active switching zone. Remember, the sensing surface is part of the oscillator in the sensor. When the rated voltage is applied to the sensor, the oscillator creates high-frequency oscillations in the active switching zone. If a metallic object is brought to this zone, eddy currents are generated in the object. The energy for generating these currents is drawn from the oscillator. As a result, the oscillator gets attenuated, triggering the switching circuit to produce an output signal. Finally, the output signal is amplified and delivered to the load connected across the corresponding terminals.

The switching distance of inductive sensors depends on the conductivity and permeability of the metal part whose presence should be detected. A reduction factor describes this dependence. The switching distance varies with the material composition of the target object, with mild steel taken as the material for standard reference. The reduction factor is the factor by which the sensing range of the inductive proximity sensor is reduced based on the material composition of the object to be sensed, compared to steel [FE 360 (St 37)] as the standard reference. The reduction factors for other materials are: 0.85 for stainless steel, 0.7–0.90 for chrome-nickel, 0.35–0.50 for aluminum and brass and 0.25–0.40 for copper.

Another factor that affects the sensing range of inductive sensors is the diameter of the sensing coil. A small sensor with a coil diameter of ¾ inch has a typical sensing range of 1 mm, while a large sensor with a coil diameter of 3 inches has a sensing range of 50 mm or more.

Inductive proximity sensors are self-contained, rugged, and extremely reliable. They can be used in a large number of applications, like: (1) sensing the end positions of a pneumatic cylinder, a semi-rotary drive, or a press ram, (2) detecting metallic workpieces on conveyors, (3) finding the speed of a rotary machine by sensing the passing gear teeth of the associated feedback device, and (4) monitoring drill breakage for fracture during the work process.

Capacitive Proximity Sensor
A capacitive proximity sensor consists of (1) an oscillator (RC resonant circuit), (2) a switching circuit, (3) an amplifier, and (4) an output stage. A capacitive proximity sensor with these blocks is shown in Figure 3.17.

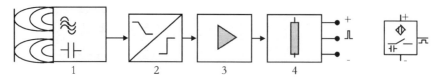

Figure 3.17 | The block diagram of a capacitive proximity sensor

The sensing surface is designed with an active electrode and an earth electrode. An electrostatic stray field is created in the active switching zone between the active and earth electrodes in front of the sensor. If an object is brought into the active switching zone, the capacitance of the resonant circuit is altered. This change in the capacitance primarily depends on the following parameters: (1) the distance of the object from the active surface, (2) the dimensions of the object, and (3) the dielectric constant of the object. The change in the capacitance of the oscillator circuit triggers the switching circuit to produce an output signal. Finally, the output signal is amplified and delivered to the load connected across the corresponding terminals.

Due to their ability to react to a wide range of materials, capacitive proximity sensors are used more universally in applications. These are suitable for detecting non-metallic objects and monitoring the filling levels of storage containers. However, capacitive proximity sensors are sensitive to the effects of humidity in the active switching zone.

Optical Proximity Sensors

An optical proximity sensor employs optical and electronic means for sensing objects. This sensor consists of two units:
- Emitter
- Receiver

The emitter is a source of infrared light rays. These rays travel in a straight line, and this property is essential for the proper sensing of objects by an optical proximity sensor. A semiconductor light-emitting diode (LED) is a particularly reliable source of infrared rays. The receiver is a photodiode or a phototransistor that can electronically accept and evaluate these infrared rays.

Two important types of the optical sensor are: (1) Through-beam sensor and (2) Diffuse sensor.

In a through-beam sensor, the emitter and receiver are mounted separately, and in a diffuse sensor, these units are mounted in a common housing.

Through-beam Sensor

This sensor consists of separate emitter and receiver units.

The emitter emits infrared rays when an electrical current is passed through it. The receiver reacts to the infrared rays, which travel straight. When no object is present in the path of the rays, they hit the receiver.

This type of proximity sensor is designed to generate an output signal only when an object interrupts these rays. A through-beam sensor with emitter and receiver units arranged in line for the emitted infrared rays to hit the receiver unit and an object approaching the path of these rays for its interruption is shown in Figure 3.18.

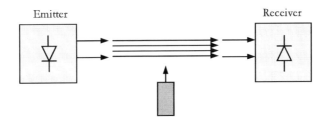

Figure 3.18 | Illustrating the principle of a through-beam sensor

The sensing distance, typically up to 2 meters (6.5 ft), if not more, is possible with different designs and makes of through-beam sensors. These sensors have a wide sensing range and good positioning accuracy. However, their disadvantage is that they need two separate proximity modules (emitter and receiver). Receivers are designed with PNP or NPN outputs. (Please see page 27 for more details)

Diffuse Sensor

In a diffuse sensor, the emitter and receiver are fitted in the same housing (Figure 3.19). In this sensor, the object diffusely reflects a percentage of the emitted light, activating the receiver.

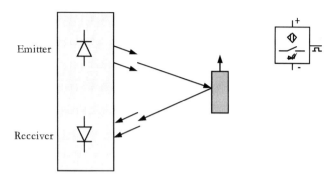

Figure 3.19 | Illustrating the working principle of a diffuse sensor

Maximum sensing distance is typically up to 160 mm (6.2 inches). The size, surface, shape, density, and color of the object determine the intensity of the diffused light emission and, thus, the actual sensing range.

Connection and Circuit Technology

As stated earlier, the three-wire proximity sensors have three connecting wires. Two wires are for the voltage supply (positive and negative). The third wire is meant for the signal output of the proximity sensor.

The output stage of a discrete proximity sensor usually consists of a transistor that acts as a switch. When the sensor detects nothing, the transistor acts as an open switch.

When the sensor detects some physical phenomenon, the transistor acts like a closed switch. Based on the connection of the transistor to the power supply terminals, discrete proximity sensors may be categorized into two types:

- Sinking output sensors
- Sourcing output sensors.

In both sinking output and sourcing output sensors, the emphasis is on the current flow to reduce electrical noise problems.

Sinking (NPN) Output Sensor

A simplified diagram of a sinking output sensor connected to an external load is given in Figure 3.20(a). The sinking output sensor has an NPN transistor output with its emitter connected to the negative potential of the supply voltage for negative switching.

The load is connected between the positive potential of the power source and the proximity sensor output.

This connection means that the sensor output will be pulled down to the negative potential, and hence the load will be connected to the negative potential through the transistor in the switched state. This connection will allow the current to flow through the sensor and then to the ground (hence sinking), as shown in Figure 3.20(a).

The advantage of this sensor is that the load can be connected to a different power source (not shown in the Figure) rather than to the same source. Therefore, this sensor is best selected when different sources of supply voltages are used for various electrical devices in the control system.

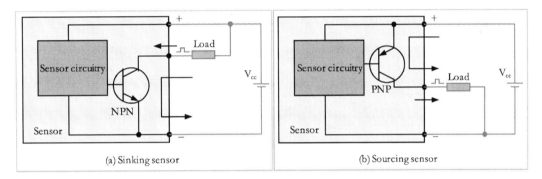

Figure 3.20 | Proximity sensor outputs

Sourcing (PNP) Output Sensor

A simplified diagram of a sourcing output sensor connected to an external load is given in Figure 3.20(b). The sourcing output sensor has a PNP transistor output with its emitter connected to the $+V_{cc}$ of the supply for positive switching.

The load is connected between the proximity sensor output and the negative potential.

This connection means that the sensor output will be pulled up to the positive potential, and hence the grounded load will be connected to the positive potential through the transistor in the switched state. This connection will allow the current to flow from the positive potential through the sensor to the output (hence sourcing), as shown in Figure 3.20(b).

This sensor is best selected when all electrical devices in the control system use a single supply voltage source. Sensors with sourcing outputs (PNP) are popular.

Example 3.8 | Semi-automatic operation of a double-acting cylinder using a 5/2-DC single-solenoid valve and limit switch

A double-acting cylinder should extend when a pushbutton is pressed. On reaching the end position, the cylinder should retract automatically. A 5/2-DC single-solenoid valve is used as the final control element. Develop an electro-pneumatic control circuit to implement the control task using a limit switch for the semi-automatic operation of the cylinder.

Solution

The double-acting cylinder is controlled by a 5/2-DC single-solenoid, spring-return valve [Figure 3.21(a)]. The electrical circuit is latched when pushbutton PB1 is pressed. Valve 1.1 remains in the actuated position even when pushbutton PB1 is released [Figure 3.21(b)]. The cylinder then starts moving forward. When the fully extended position of the cylinder is reached, it automatically actuates limit switch S2. Actuation of the limit switch causes interruption of the electrical circuit [Figure 3.21(c)]. Valve 1.1 returns to its normal position. The cylinder then retracts automatically.

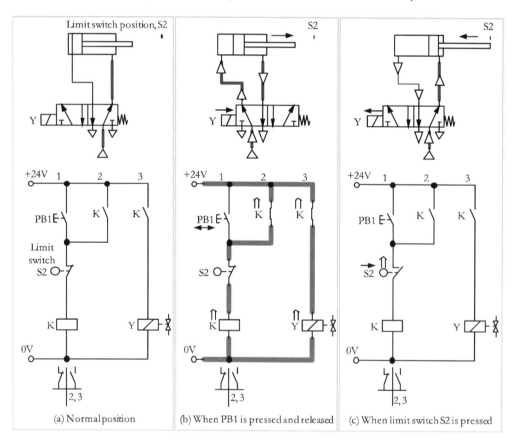

Figure 3.21 | Auto-return of the double-acting cylinder using a limit switch (Example 3.8)

Example 3.9 | Semi-automatic operation of a double-acting cylinder using a 5/2-DC single-solenoid valve and proximity sensor

A double-acting cylinder should extend when a pushbutton is pressed. On reaching the end position, the cylinder should retract automatically. A 5/2-DC single-solenoid valve is used as the final control element. Develop an electro-pneumatic control circuit to implement the control task using a proximity sensor for the semi-automatic operation of the cylinder.

Solution

The double-acting cylinder is controlled by a 5/2-DC single-solenoid, spring-return valve [Figure 3.22(a)].

The position of the circuit, when pushbutton PB1 is pressed and then released, is given in Figure 3.22(b).

The cylinder automatically extends to its forward-end position, influencing the proximity sensor S2. Relay K2 is connected across the proximity sensor to convert the voltage output of the proximity sensor to the corresponding contact operation. The normally-closed (NC) contact of relay K2 used in section 1 interrupts the latching circuit when the sensor is sensing. This position of the circuit is shown in Figure 3.22 (c). Valve 1.1 returns to its normal position. The cylinder then retracts automatically.

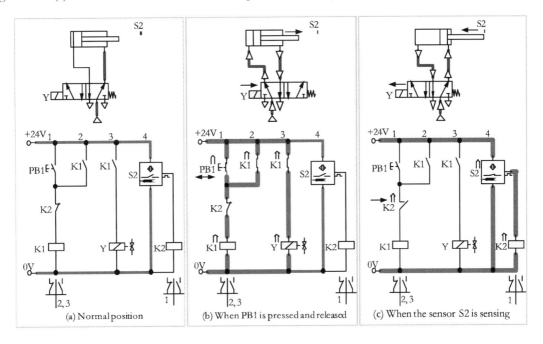

Figure 3.22 | Auto-return of the double-acting cylinder using a proximity sensor (Example 3.9)

Example 3.10 | Semi-automatic operation of a double-acting cylinder using 5/2-DC double-solenoid valve and limit switch

A double-acting cylinder should extend when a pushbutton is pressed. On reaching the end position, the cylinder should retract automatically. A 5/2-DC double-solenoid valve is used as the final control element. Develop an electro-pneumatic control circuit to implement the control task using a limit switch for the semi-automatic operation of the cylinder.

Solution
The double-acting cylinder is controlled by the 5/2-DC double-solenoid valve with solenoids Y1 and Y2 [Figure 3.23(a)].

The solenoid Y1 is energized through relay K1 when pushbutton PB1 is pressed momentarily, as shown in Figure 3.23(b). The solenoid valve is actuated to its left envelope and remains in that position even when pushbutton PB1 is released. The cylinder then extends.

When the fully extended position of the cylinder is reached, it automatically actuates limit switch S2, as shown in Figure 3.23(c). Solenoid Y2 is energized through relay K2 when the limit switch is actuated. The solenoid valve is actuated to its right envelope and remains in that position even when limit switch S2 is released. The cylinder then retracts automatically.

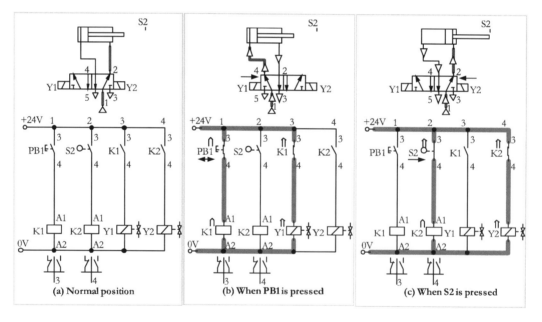

Figure 3.23 | Auto-return of the double-acting cylinder using a limit switch (Example 3.10)

Example 3.11 | Semi-automatic operation of a double-acting cylinder using a 5/2-DC double-solenoid valve and proximity sensor

A double-acting cylinder should extend when a pushbutton is pressed. On reaching the end position, the cylinder should retract automatically. A 5/2-DC double-solenoid valve is used as the final control element. Develop an electro-pneumatic control circuit to implement the control task using a proximity sensor for the semi-automatic operation of the cylinder.

Solution

The double-acting cylinder is controlled by the 5/2-DC double-solenoid valve with solenoids Y1 and Y2 [Figure 3.24(a)].

Solenoid Y1 is energized when pushbutton PB1 is pressed momentarily, as shown in Figure 3.24(b). The solenoid valve is actuated to its left envelope and remains in that position even when pushbutton PB1 is released. The cylinder then extends.

When the fully extended position of the cylinder is reached, it automatically activates proximity sensor S2, as shown in Figure 3.24(c). The solenoid valve is actuated to its right envelope and remains in that position even when sensor S2 is released. The cylinder then retracts automatically.

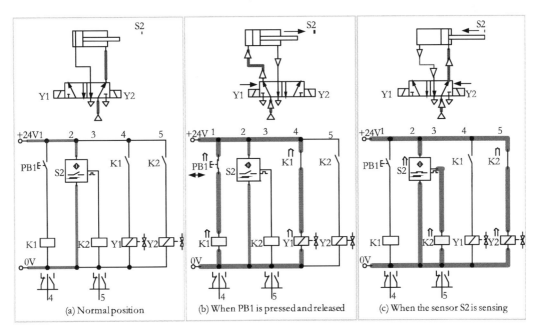

Figure 3.24 | Auto-return of the double-acting cylinder using a proximity sensor (Example 3.11)

Time-delay Relays (Timers)

Timers are required in control systems to effect time delays between work operations. This control is possible by delaying the operation of the associated control element through a timer. Nowadays, electronic timers are the most popular. Mainly, an electronic timer consists of a relay with the addition of an electronic circuit which delays the action of its control contacts. The delay time can be set on the timer using a potentiometer. The contact operation can be delayed when the coil is energized or de-energized. Accordingly, there are two basic types of timers: (i) an on-delay timer and (ii) an off-delay timer. The symbols of these types of timers are given in Figure 3.25(a) and 3.25(b), respectively. The operations of these types of timers are explained below.

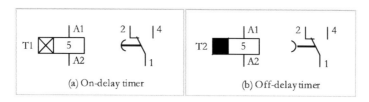

Figure 3.25 | Symbolic representations of time-delay relays

In the on-delay timer, shown in Figure 3.26(a), when the pushbutton is pressed (ON), capacitor C is charged through potentiometer R_1 as diode D is reverse-biased. The time taken to charge the capacitor depends on the resistance of the potentiometer (R_1) and the capacitance (C) of the capacitor. The required time delay can be set by adjusting the potentiometer's resistance. Coil K is energized when the capacitor is charged sufficiently, and its contacts are operated after the set time delay.

When the pushbutton is released (OFF), the capacitor discharges quickly through a small resistance (R_2) as the diode bypasses resistor R_1 and the contacts of relay K return to their normal position without any delay.

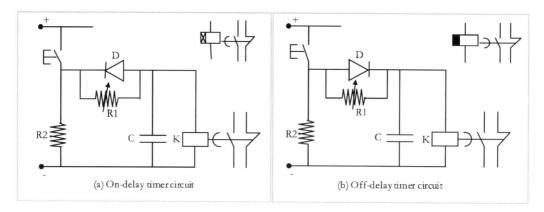

Figure 3.26 | Simplified circuits of electronic timers

In the off-delay timer, shown in Figure 3.26(b), the contacts are operated without delay when the pushbutton is pressed (ON). When the pushbutton is released (OFF), the contacts return to the normal position after the set delay.

Example 3.12 | Control of double-acting cylinder using a timer

A double-acting cylinder should extend when a pushbutton is pressed (short pulse). It is to remain in the extended position for 5 seconds and then return automatically. The final forward position of the cylinder is registered with a proximity sensor S2. A 5/2-DC double-solenoid valve is used as the final control element. Develop an electro-pneumatic control circuit to implement the control task.

Solution

A latching circuit is used to obtain the necessary memory function. The position of the circuit, when pushbutton PB1 is pressed and then released, is given in Figure 3.27(a). The cylinder extends to its forward-end position and actuates limit switch S2 automatically.

As the return motion is to be delayed, an on-delay timer is used to obtain the necessary time delay. The required time delay should be set on the timer. Limit switch S2 controls the timer coil T. After the set delay, the timer contact interrupts the latching circuit, thus causing the return motion of the cylinder, as shown in Figure 3.27(b).

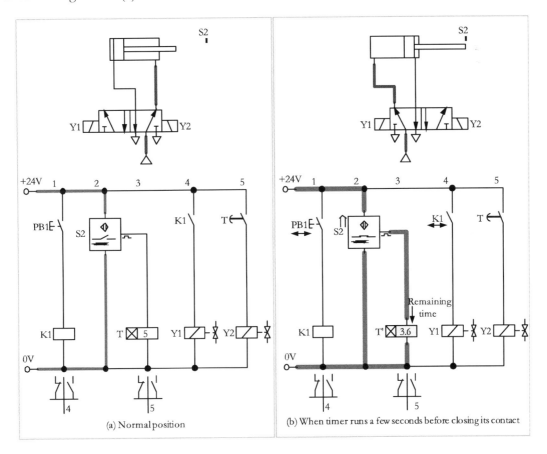

Figure 3.27 | The control of a double-acting cylinder using a timer
(Example 3.12)

Example 3.13 | Control of double-acting cylinder using a timer

Control task: A double-acting cylinder should extend when a pushbutton is pressed (short pulse). It is to remain in the extended position for 5 seconds and then return automatically. The final forward position of the cylinder is registered with a limit switch S2. A 5/2-DC single-solenoid valve is used as the final control element. Develop an electro-pneumatic control circuit to implement the control task.

Solution

A latching circuit is used to obtain the necessary memory function. The position of the circuit, when pushbutton PB1 is pressed and then released, is given in Figure 3.28(a). The cylinder extends to its forward-end position and actuates limit switch S2 automatically.

As the return motion is to be delayed, an on-delay timer is used to obtain the necessary time delay. The required time delay should be set on the timer. Limit switch S2 controls timer coil T. After the set delay, the timer contact interrupts the latching circuit, thus causing the return motion of the cylinder, as shown in Figure 3.28(b).

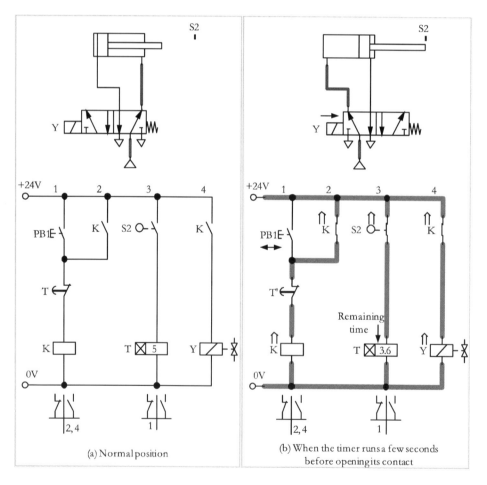

Figure 3.28 | The control of a double-acting cylinder using a timer
(Example 3.13)

Two-hand Safety Operation
A machine for pressing, cutting, or other similar operations is usually designed with a two-hand safety feature using two pushbuttons (PBs) installed so that both cannot be pressed simultaneously by one hand. An operator can operate the machine only by pressing the two pushbuttons simultaneously. That means the operator's both hands must be used to operate the machine. This feature ensures that the operator's hand cannot be on the machine while operating for safety reasons.

Further, the machine should not operate when one pushbutton is tied down permanently by some means. An operator may do this trick to free one hand to adjust the active workpiece while the machine operates. This way of working is hazardous. Anti-tie-down and anti-repeat circuits can ensure that both switches must be OFF and pressed simultaneously within a short duration, usually within half a second, to operate the machine for one cycle. The pushbuttons must be released to start the next cycle, and the entire process must be repeated. This control task is given as an assignment [A3.2 (Page 40)] for the reader to attempt.

Pressure Switch
A pressure switch is a pneumatic–electric (P/E) signal converter. These switches fall into three general classes. In the first type of pressure switch design, bellows are used, which expand or contract in response to an increase or decrease in pressure. The contacts are mounted on the end of a lever which is acted upon by the bellows.

In the second type, a diaphragm is used instead of bellows, as shown in Figure 3.29. When the compressed air is applied at the inlet, and the pre-set pressure is reached, the diaphragm expands and pushes the spring-loaded plunger. This force, in turn, operates the contacts.

In the third type of design, a Bourdon tube is used to actuate the contacts through a specific mechanism. The Bourdon tube employs a hollow tube in a semi-circular shape whose design is such that an increase in pressure tends to straighten it. The resulting force is sufficient to actuate an integrated snap-action switch. The pressure is adjustable in the range from 1 to 10 bar.

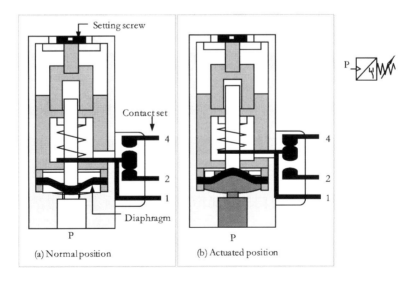

Figure 3.29 | A pressure switch

Example 3.14 | Stamping device

Components are to be stamped using a stamping device. A double-acting cylinder pushes the die attached to a fixture when a pushbutton (PB1) is pressed. The die is to return to the initial position upon reaching sufficient stamping pressure. A pressure switch senses the stamping pressure. Develop an electro-pneumatic control circuit to implement the control task for the stamping operation.

Solution

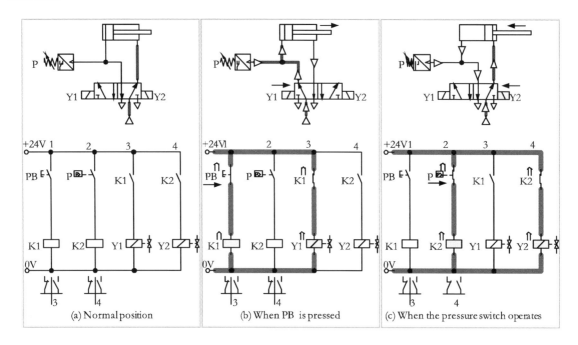

Figure 3.30 | Electro-pneumatic circuit for the stamping device
(Example 3.14)

The double-acting cylinder is controlled by a 5/2 DC valve, as shown in Figure 3.30(a). A pressure switch is connected to the line leading to the piston side of the cylinder.

When pushbutton PB is pressed, solenoid coil Y1 is energized through relay K1, as shown in Figure 3.30(b). The DC valve switches over, and the cylinder extends to the forward-end position.

When the pre-set switching pressure is reached in the supply line of the cylinder piston side, pressure switch P is activated, as shown in Figure 3.30(c). Consequently, relay coil K2 and solenoid coil Y2 are energized. The DC valve switches back to its start position, and then the cylinder returns to its rear-end position.

Example 3.15 | Cyclic operation of a double-acting cylinder

When a 'Start' pushbutton is pressed, a double-acting cylinder should perform a continuous back-and-forth motion until a 'Stop' pushbutton is pressed. The cylinder should stop in the retracted position always. A 5/2-DC double solenoid valve is used as the final control element. Develop an electro-pneumatic control circuit for implementing the fully automatic operation of the cylinder.

Solution

The electro-pneumatic circuit for the cyclic operation of a double-acting cylinder controlled by a 5/2-double-solenoid valve is given in Figure 3.31(a). Limit switches S1 and S2 are positioned for actuation by the cylinder at the retracted and extended positions. Limit switch S1 is actuated in the initial position. This actuated condition of the switch is represented in the drawing with an arrow alongside.

The fully automatic cyclic operation of the cylinder can be obtained simply by using sensor signal S1 controlling coil Y1 through relay coil K2 and sensor signal S2 controlling coil Y2 through relay coil K3.

The 'start' and 'stop' controls of the cyclic operation can be incorporated by using a latching circuit controlled by pushbuttons PB1 (Start) and PB2 (Stop). A contact of relay coil K1 is connected in series with the sensor S1 contact in branch 3 to obtain the necessary 'start' and 'stop' controls. The position of the circuit when proximity sensor S2 is activated is shown in Figure 3.31(b).

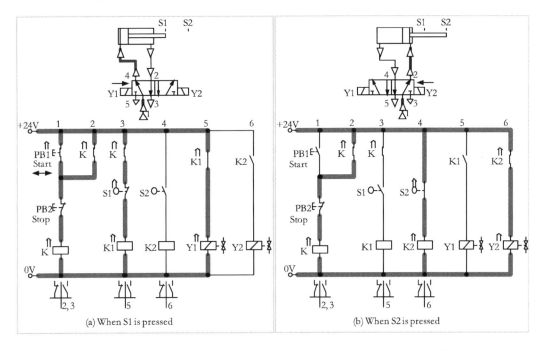

Figure 3.31 | Circuit for the cyclic operation of a double-acting cylinder
(Example 3.15)

Preset Counters

An electrically operated counter consists of a coil, associated circuits and contacts, a reset coil, a manual reset, a release button, and a display window. Pressing the release button of the counter and entering the desired count value set the pre-determining counter. The pre-determined count value is displayed in the window. An up counter counts electrical signals upwards from zero. For each electrical counting pulse input to an up-counter coil, the count value is incremented by 1. When the preset value has been reached, the relay picks up, and the contact set is actuated. A down counter counts electrical signals downwards from a preset number. If the count value of zero is reached, the relay picks up, and the contact set is actuated. The counter can be reset manually by pressing the reset button or electrically by applying a pulse to the reset coil. The pre-determined value is maintained when the counter is reset. The symbol of an up counter is given in Figure 3.32.

Figure 3.32 | Symbol of a counter, electric

Example 3.16 | Control circuit using a counter

When a 'Start' pushbutton is pressed, a double-acting cylinder is to perform a continuous back-and-forth motion. The cylinder should stop automatically after performing 5 cycles of operation. Develop an electro-pneumatic control circuit using a down counter.

Solution

The electro-pneumatic circuit for the cyclic operation of a double-acting cylinder controlled by a 5/2-DC double-solenoid valve is given in Figure 3.33(a). Limit switches S1 and S2 are positioned for actuation by the cylinder at the retracted and extended positions.

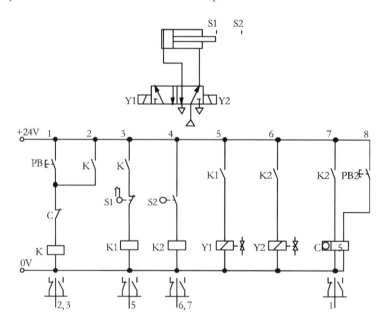

Figure 3.33(a) | The automatic stopping of cyclic operation of a cylinder (Example 3.16)

Solenoid coil Y1 controls the forward motion of the cylinder through the contact of relay K1, which is controlled by the contact of relay K and the contact of sensor S1 in a series connection. Solenoid Y2 controls the return motion through a contact of relay K2 which is, in turn, controlled by the contact of sensor S2. In each cycle, a signal pulse from sensor S2 is input to the counter coil through relay K2. The NC contact of the counter is used to interrupt the latching circuit and stop the cyclic operation after the set number of cycles of operation is performed. The position of the circuit when the counter coil receives a countdown signal is given in Figure 3.33(b). The position when counter contact C interrupts the latching circuit to stop the cyclic operation is given in Figure 3.33(c).

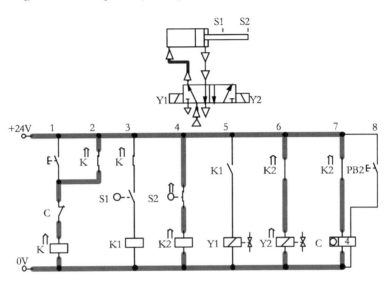

Figure 3.33(b) | The position when the counter coil A1-A2 is receiving a countdown signal

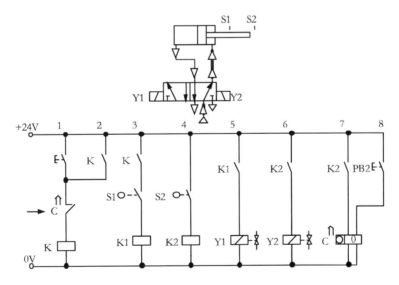

Figure 3.33(c) | The position when the counter contact C interrupts the latching circuit to stop the cyclic operation

Assignments – Basic Level

A3.1 | A pneumatically controlled press with a stamping die, as shown in Figure 3.34, produces badges from a very thin metal sheet. A safety guard is provided for the safety of the operator. A double-acting cylinder is used as the drive for the press. The cylinder should extend when the safety guard is placed, and pushbutton PB1 is pressed. The cylinder should retract automatically after reaching the forward end position and attaining a preset pressure for consistent quality. The cylinder should retract immediately if the emergency push-button PB2 is pressed. Develop an electro-pneumatic control circuit to implement the control task.

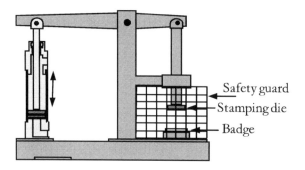

Figure 3.34 | A stamping die

A3.2 | A double-acting cylinder embosses slide rules, as shown in Figure 3.35. The cylinder extends when only two push-buttons are pressed simultaneously within one second to ensure the complete safety of operators. The cylinder should retract immediately if either one or both pushbuttons are released. Develop a pneumatic circuit to implement the control task.

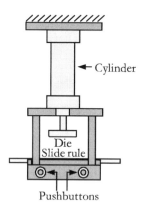

Figure 3.35 | An embossing machine

A3.3 | Figure 3.36 shows the arrangement for cleaning washers for injection pumps in a cleaning bath. A double-acting cylinder moves the container filled with washers up and down in the bath several times. The operator provides a 'Start' signal manually. The washing operation is turned off automatically after a preset time. Develop a pneumatic circuit to implement the control task.

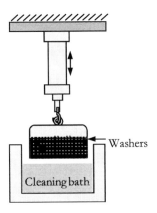

Figure 3.36 | A Cleaning bath for washers

A3.4 | An arrangement for pressing components for 20 seconds is shown in Figure 3.37. A push-button PB1 is used to control the forward stroke of the cylinder. After pressing the components for 20 seconds, the cylinder should retract automatically. The return stroke must occur even if the start pushbutton is still depressed. A new start signal may only be effective after the initial position of the piston is reached, and the pushbutton is released. Develop a pneumatic circuit to implement the control task.

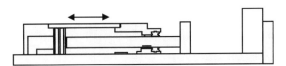

Figure 3.37 | A Cleaning bath for washers

Solution - Assignment No. A3.1 (Pg. 40)

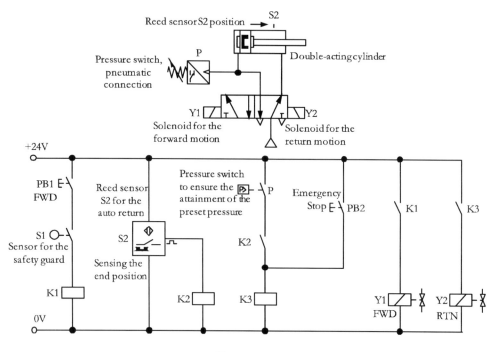

Figure 3.38

Solution - Assignment No. A3.2 (Pg. 40)

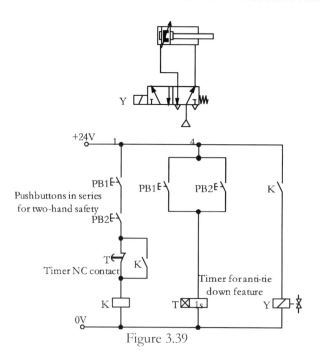

Figure 3.39

Solution - Assignment No. A3.3 (Pg. 41)

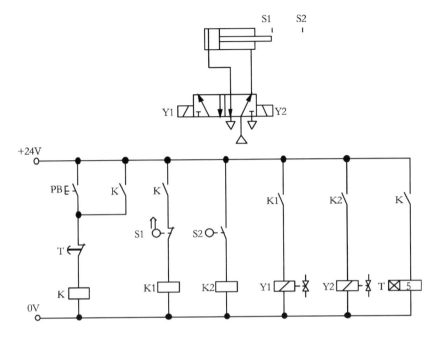

Figure 3.40

Solution - Assignment No. A3.4 (Pg. 41)

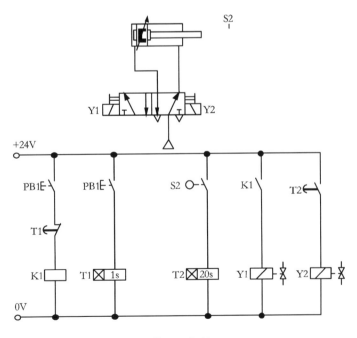

Figure 3.41

Chapter 4 | Multiple-actuator Electro-pneumatic Circuits

The concepts of developing basic electro-pneumatic circuits involving only one cylinder were discussed in Chapter 3. The orderly way of developing electro-pneumatic circuits should be applied to more extensive multiple-actuator electro-pneumatic systems. It may be remembered that double-solenoid valves are most commonly employed in electro-pneumatic systems to control pneumatic cylinders due to their most useful memory characteristic. However, a double solenoid valve is susceptible to the problem of signal conflict or signal overlap, which is due to the appearance of signals at both solenoids simultaneously. The apparent result of signal conflict is that the valve will not operate properly. Therefore, the main requirement in developing multiple-actuator electro-pneumatic circuits is the knowledge of different ways to eliminate signal conflicts.

Representation of a Control Task 4.1
Many concepts in the development of electro-pneumatic systems are explained with the help of a problem or control task. A control task can be expressed in text form, positional layout, notational form, or displacement-step diagram. The following sections present the representations of a control task in many ways.

Text Form
Cylinders A(1.0) and cylinder B(2.0) are used to carry out stamping operations. Cylinder A extends and brings a workpiece under the stamping cylinder B. Cylinder B then extends and stamps the workpiece. Cylinder A can return only after cylinder B has retracted fully. An automatic electro-pneumatic sequential circuit must be developed to realize the control task.

Positional Layout
The arrangement of pneumatic cylinders for the control task governing the stamping operation can be expressed through the positional sketch in Figure 4.1.

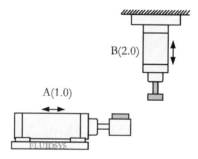

Figure 4.1 | The arrangement of pneumatic cylinders for the stamping operation

The required sequence can be stated as follows: First, Cylinder A moves forward, bringing the workpiece under the stamping cylinder B, and then Cylinder B moves forward and carries out the stamping operation. Next, Cylinder B retracts fully, and then Cylinder A can retract.

Notational Form

In the notational form, cylinders are designated A, B, C, etc. A '+' sign represents the forward stroke of a cylinder. A '-' sign represents the return stroke of a cylinder. For example,

A+ represents the forward motion of cylinder A
A- represents the return motion of cylinder A
B+ represents the forward motion of cylinder B
B- represents the return motion of cylinder B

Therefore, the notational form of representation for control task 4.1 given on Page 41 can be written as **A+ B+ B- A-**.

Displacement-Step Diagram

The displacement-step diagram for control task 4.1 is given in Figure 4.2. In this way of representation, the displacements of cylinders are plotted according to the required sequence of cylinder actions in equal steps. As shown in the Figure, a cylinder can move from the retracted position (0) to the extended position (1) and vice versa.

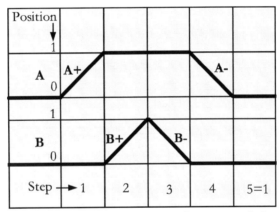

Cylinder Position: 0 – Retracted, 1 – Extended

Figure 4.2 | Displacement – Step diagram

Circuit Design for the Sequence A+ B+ B- A- [Control Task #4.1]

Sensors S1 and S2 are used for sensing the retracted and extended positions of Cylinder A, respectively. Sensors S3 and S4 are used for sensing the retracted and extended positions of Cylinder B, respectively. Pushbuttons for the start, stop, and emergency stop controls may also be necessary as per the requirement.

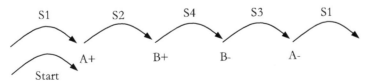

Figure 4.3 | Notational form

The first step in drafting a circuit is to write down the notations for the desired sequence of cylinder actions and the associated sensor designations, as shown in Figure 4.3. First, draw the pneumatic power circuit [Figure 4.4(a)], and then draw the relay control circuit [Figure 4.4(b)]. The developed circuit must be checked for the presence of signal conflicts.

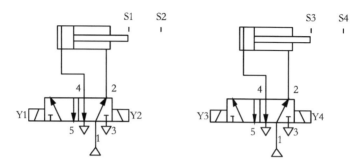

Figure 4.4(a) | Pneumatic part of the circuit

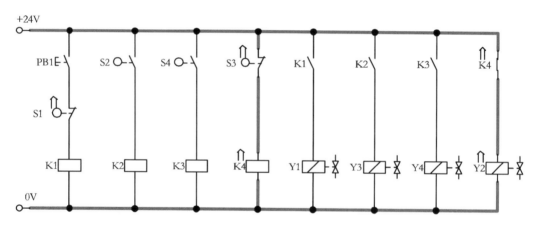

Figure 4.4(b) | Electrical part of the circuit in the initial position

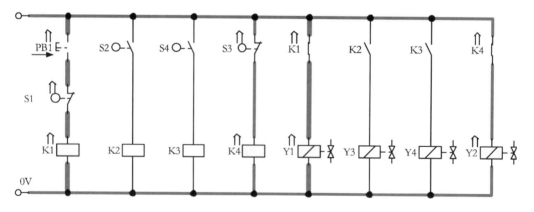

Figure 4.4(c) | Electrical part of the circuit when PB1 is pressed

When pushbutton PB1 is pressed and sensor S1 is actuated, solenoid Y1 gets energized through relay K1, as shown in Figure 4.4(c). At the same time, solenoid Y2 is also energized through relay K4, as sensor S3 remains actuated. The simultaneous energization of solenoids Y1 and Y2 of valve 1.1 results in a signal conflict, and the valve cannot switch over.

Elimination of Signal Conflicts

Various methods are devised to solve the problem of signal conflicts in multi-cylinder electro-pneumatic circuits. A commonly used method uses relays to control the power supply to different sections of the control circuit.

In this method, the sequence of operations of cylinders (say, A+B+B-A-) can be divided into an appropriate number of groups so there is no possibility of signal conflicts. If A+ and A- operations are in one group, signals can appear simultaneously at both ends (Y1 and Y2) of the solenoid valve controlling cylinder A, which is a signal conflict state. Similar is the case for B+ and B- operations. Hence the sequence of operations is divided so that cylinder actions A+ and A- fall into different groups, cylinder actions B+ and B- fall into different groups, and so on. Such a grouping is shown in Figure 4.5.

Figure 4.5 | Grouping of a sequence of operations

As shown in the Figure, the first two cylinder actions A+ and B+ can be placed in group G1. The next two cylinder actions B- and A-, if placed in group G1, may cause signal conflicts and hence are placed in group G2.

Remember, the desired sequence of operations should be retained. Another factor that should be remembered is that as the number of groups increases, the number of relays also increases. Hence, every attempt should be made to keep the number of groups minimum.

After dividing the sequence of cylinder actions into many groups, the next requirement is to divide the electric power supply into the same number of groups as cylinder actions. The grouping must be in such a way that at any given point in time, only one group will have a supply, with all other groups disconnected from the supply.

A two-group power supply can quickly be developed using the NO and NC contacts of a relay. However, for more than two supply groups, the number of relays will be the same as the number of power supply groups. Each power supply group is controlled by a NO contact of the respective relay in a sequential manner.

Design for a Two-group Electro-pneumatic Circuit

In developing a simple two-group electro-pneumatic circuit, it is necessary to divide the power supply into two groups so that at any point, only one group is live with the other group switched off. A two-group circuit can efficiently be designed using a single relay. The structure of a group-changing cascade circuit for two groups (say G1 and G2) using a relay is given in Figure 4.6.

The first part of the structure is the arrangement of the simple latching circuit with blocks for 'Signals for the 1st Group', 'Signal for the 2nd group', 'Relay', and 'Latching contact', as shown in Figure 4.6.

The block 'Signals for the 1st Group' represents all control contacts for changing the group from G2 to G1.

The block 'Signals for the 2nd Group' represents the control contacts for changing the group from G1 to G2.

The blocks 'Relay' and 'Latching contact' represent the relay coil and its contact used to obtain the cascade operation's required memory function.

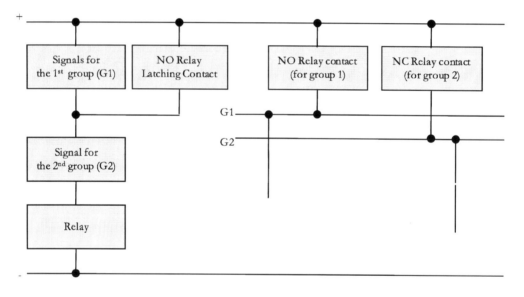

Figure 4.6 | Structure of group-changing relay circuit for two groups

The second part of the structure consists of two blocks – 'NO contact' and 'NC contact'. These blocks represent the NO and NC contact of the relay to switch groups G2 to G1 and G1 to G2, respectively.

It can be observed that initially, group G2 is live through the NC contact of the relay, as shown in the Figure. This group-changing concept is implemented in the circuit given in Figure 4.8.

Example 4.1 | Pneumatically-controlled stamping device
Cylinder A (1.0) extends and brings a job under the stamping cylinder B (2.0). Cylinder B then extends and stamps the job. Cylinder A can return only after cylinder B has retracted fully. An electro-pneumatic control circuit has to be developed to realize the control task.

Solution
The notational form representing the given control task in Example 4.1 and the grouping that will eliminate the possibility of signal conflicts are given in Figure 4.7. An electro-pneumatic control circuit is also developed with usual notations, as shown in Figure 4.8(a).

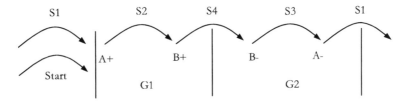

Figure 4.7 | Notational form of representation of the pneumatically-controlled stamping device

Next, divide the electrical power supply into an equal number of groups so that only one group remains energized with all other groups remaining switched off.

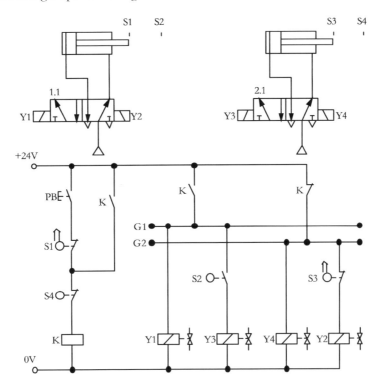

Figure 4.8(a) | Electro-pneumatic circuit for the control task of the stamping device
(Example 4.1)

Initially, the supply is in the last group, G2. The sequence of actions when pushbutton PB is pressed is explained below. When pushbutton PB is pressed, relay coil K is energized, and the circuit is latched by an NO contact of relay K in branch 2. In the main circuit, group G1 is live through another NO contact of relay K, and group G2 is switched off. Solenoid coil Y1 is energized; solenoid coil Y2 is de-energized; hence, valve 1.1 experiences no signal conflict. Valve 1.1 changes its state, and cylinder A travels out (A+).

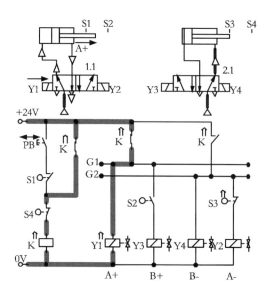

Figure 4.8(b) | A+

The limit switch S2 gets actuated, thereby energizing solenoid coil Y3. As a result, valve 2.1 switches over, and cylinder B travels out (B+). Limit switch S4 gets actuated now.

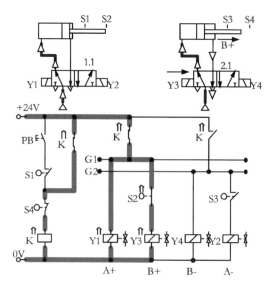

Figure 4.8(c) | B+

The signal from sensor S4 must perform a signal shut-off, and hence its NC contact is incorporated in branch 1 of the circuit. The latched circuit of relay coil K is cleared by the actuated limit switch S4. In the main circuit, group G1 is again disconnected through the released NO contact of relay K in branch 3, and group G2 is resupplied with the current through the NC contact of relay K in branch 4.

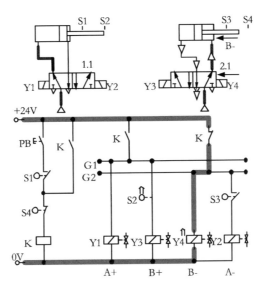

Figure 4.8(d) | B-

Solenoid coil Y4 is energized, and solenoid coil Y3 is de-energized; hence valve 2.1 experiences no signal conflict. Valve 2.1 changes its state, and Cylinder B returns to the rear-end position (B-). Limit switch S3 is actuated, thereby energizing solenoid coil Y2. Valve 1.1 switches over, and Cylinder A returns to the rear-end position (A-). The cycle of cylinder actions can be repeated.

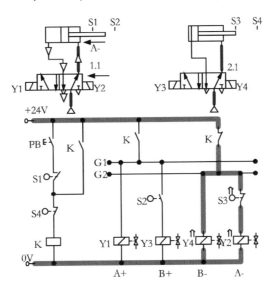

Figure 4.8(e) | A-

Design for a Three-group Electro-pneumatic Circuit

In developing electro-pneumatic circuits with three or more groups, it is necessary to divide the power supply into three or more groups so that at any point in time, only one group is live with all other groups switched off. A three-group circuit can be designed using three relays. This technique is generally called the shift register method. The general structure of a group-changing cascade circuit for 'n' groups (i.e., groups G1, G2... Gn) using as many relays as the number of groups is given in Figure 4.9.

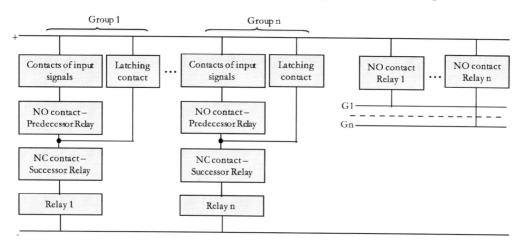

Figure 4.9 | General structure of group changing circuits using relays

The structure is divided into many identical layouts such as Group_1, Group_2, …. Group_n. Normally-open (NO) contacts of Relay_1 to Relay_n are used to switch the group power supply from G1 to Gn, respectively. The Group_1 part controls the Group G1 supply through Relay_1; the Group_2 part controls the Group G2 supply, and so on.

The concept of the group-changing circuit is explained by taking the Group_1 part of the structure. It consists of the following blocks: (1) 'Contacts of input signals', (2) 'NO contact-Predecessor Relay', (3) 'NC contact-Successor Relay', (4) 'Relay_1', and (5) 'Latching contact'.

The blocks 'Input signals' and 'NO contact-Predecessor Relay' represent all control contacts in the series connection necessary for changing the group from the last group (say, Gn) to G1. The block 'NO contact-Predecessor Relay' represents the NO contact of the predecessor relay. It sets the cascade operation in the required sequence (i.e., G1, G2 …. Gn). That means a particular cascade group can be set only if the predecessor group is already set. The block 'NC contact-Successor Relay' represents the control contact used to reset the group immediately after the successor group is set. The block 'Relay Coil 1' is the relay coil in group 1, and the 'Latching contact' represents the NO contact used for obtaining the required latching of the group.

The final part of the structure consists of blocks 'NO contact Relay 1', 'NO contact Relay 2' etc. These blocks represent the NO contacts of Relay 1, and Relay 2, respectively. These contacts switch the respective supply groups G1, G2, etc. The last group should initially be set as usual in pneumatic/electro-pneumatic systems. The shift register circuit then changes the group in sequences G1 and G2… Gn in response to appropriate 'Input signals'.

Example 4.2 | Pneumatically-controlled drilling machine

Workpieces are to be drilled using a pneumatically-controlled drilling machine given in Figure 4.10. The workpieces are arranged in a gravity-feed magazine. The workpieces are pushed and clamped using a clamping cylinder A, drilled by a drilling cylinder B, and ejected by an ejecting cylinder C. Develop an electro-pneumatic control circuit to implement the control task as given in the associated displacement-step diagram.

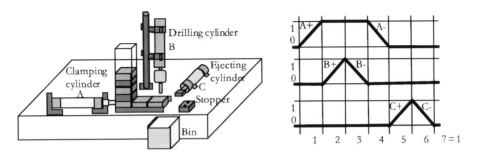

Figure 4.10 | Pneumatically-controlled drilling machine (Example 4.2)

Solution

The notational form to represent the control task of Example 4.2, along with their grouping, is given in Figure 4.11. The details of the resulting sensor signals are also given in the Figure.

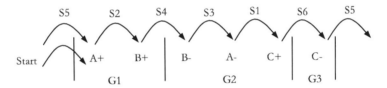

Figure 4.11 | Notational form for the control task (Example 4.2)

Cylinder Actions are divided into groups in such a way that there is no possibility of signal conflicts. The given sequence of operations is divided into three groups, as shown in the figure.

The control circuit, given in Figure 4.12(a) for dividing the electrical power supply into three groups, is designed as explained in the earlier section. Three relays (K1, K2, and K3) are used along with other control contacts to construct the group circuits. The NO contacts of relays K1, K2, and K3 divide and control the electrical power supply into three groups (G1, G2, and G3). The complete circuit to implement the control task of Example 4.2 is given in Figure 4.12(a).

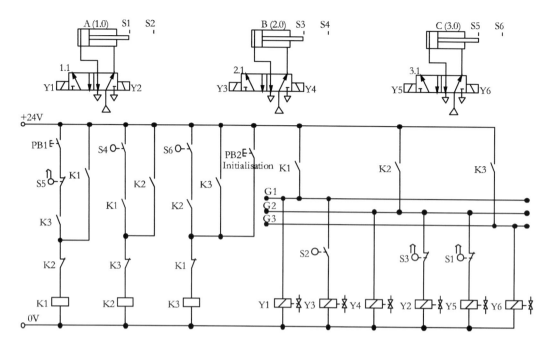

Figure 4.12(a) | Electro-pneumatic circuit for the control task of drilling machine
(Example 4.2)

If the control system is being commissioned for the first time or if there has been a power failure, the control system must be initialized with the last group (G3) connected to the power supply and all other groups disconnected from the power supply. For this reason, an initialization pushbutton (PB2) is incorporated into the circuit of relay coil K3 [Figure 4.12(a)].

The actuation of the 'Start' pushbutton PB1 results in the latching of relay K1 and the consequent unlatching of relay K3, as shown in Figure 4.12(b). Group G1 is now connected to the power supply through relay contact K1. All other groups (G2 and G3) remain disconnected from the power supply. In group G1, solenoid coils Y1 and Y3 are energized sequentially through sensor S2, resulting in cylinder actions A+ and B+, as illustrated in Figures 4.12(b) and 4.12(c).

As Cylinder B moves forward, sensor S4 is activated. The actuation of sensor S4 results in the latching of relay K2 and the consequent unlatching of relay K1, as shown in Figure 4.12(d). Group G2 is connected to the power supply through relay contact K2. All other groups (G1 and G3) remain disconnected from the power supply. In group G2, coils Y4, Y2, and Y5 are energized in sequence through sensors S3 and S1, resulting in cylinder actions B-, A-, and C+, as illustrated in Figures 4.12(d), 4.12(e), and 4.12(f).

As cylinder C moves forward, sensor S6 is activated. The actuation of sensor S6 results in the latching of relay K3 and the consequent unlatching of relay K2, as shown in Figure 4.12(g). Group G3 is connected to the power supply through the relay contact K3. All other groups (G1 and G2) remain disconnected from the power supply. In group G3, coil Y6 is energized, resulting in cylinder action C-, as illustrated in Figure 4.12(g).

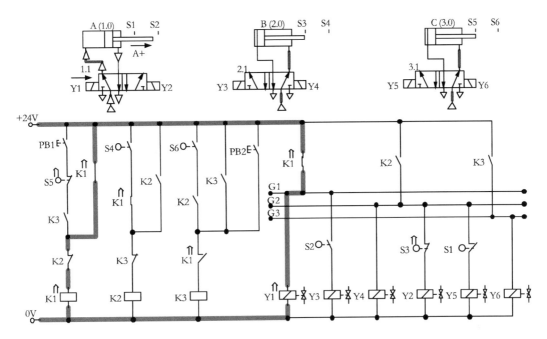

Figure 4.12(b) | A+

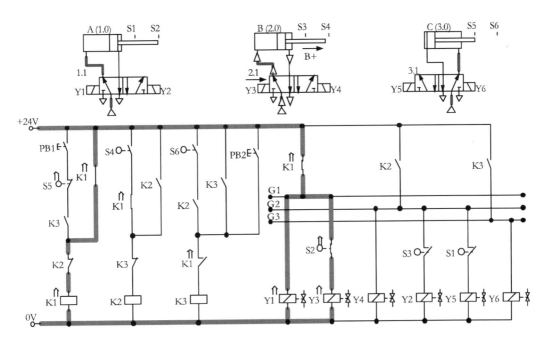

Figure 4.12(c) | B+

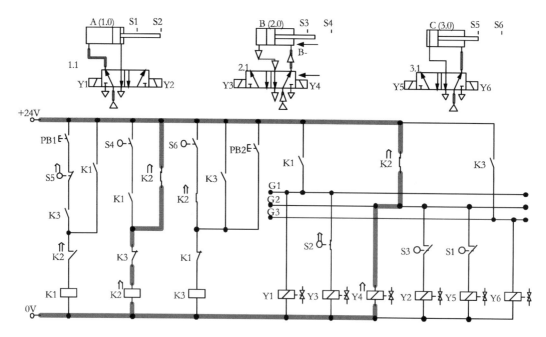

Figure 4.12(d) | B-

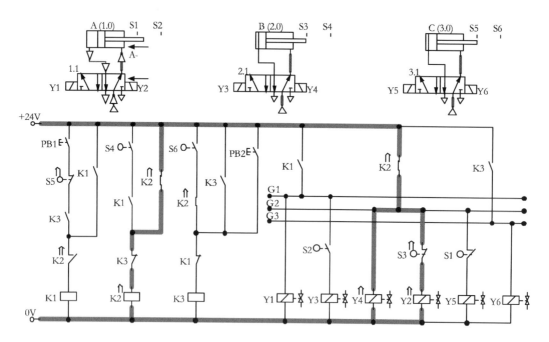

Figure 4.12(e) | A-

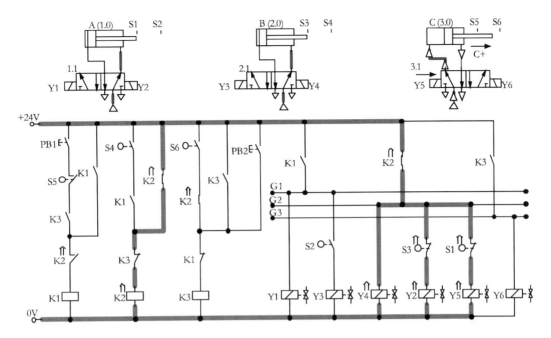

Figure 4.12(f) | C+

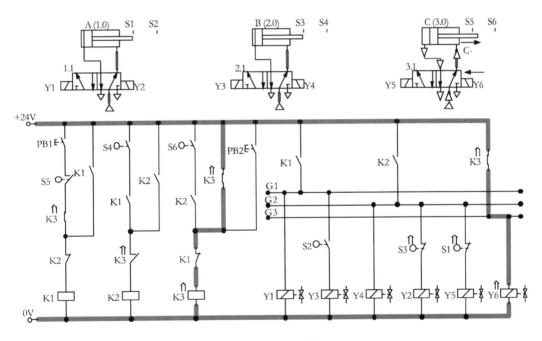

Figure 4.12(g) | C-

Assignments – Advanced Level

Problem No. 1 | Pneumatically-actuated Lathe

The arrangement for an automatically operated pneumatic lathe is shown in Figure 4.13(a). First, cylinder 1.0 shifts the entire feed magazine to the transfer station. Then, cylinder 2.0 pushes the workpiece into the lathe chuck. Cylinder 1.0 and 2.0, then retract. Cylinder 3.0 moves the slide rest forward and back. Cylinder 4.0 then ejects the workpiece following the release of the lathe chuck. The sequence of operations of these cylinders is shown in the displacement-step diagram [Figure 4.13(b)]. Develop a pneumatic control circuit to implement the given control task: (i) for one cycle and (ii) for a continuous sequence of operations.

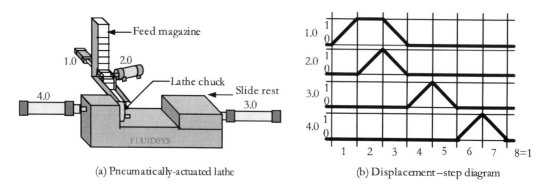

(a) Pneumatically-actuated lathe   (b) Displacement–step diagram

Figure 4.13 | Pneumatically-actuated lathe

Problem No. 2 | Processing of Workpieces

An arrangement for the 'feeding, clamping, processing, and ejecting' of workpieces is shown in Figure 4.14(a). Workpieces are taken from the gravity feed magazine and placed into the device by cylinder 1.0. Next, cylinder 2.0 extends and clamps the workpiece. Cylinder 3.0 is used to carry out the processing. The processing time is adjustable via a timer. Cylinder 4.0 is used to eject the workpiece. The required sequence of operation is shown in the sequence diagram [Figure 4.14(b)]. Develop a pneumatic control circuit to implement the control task.

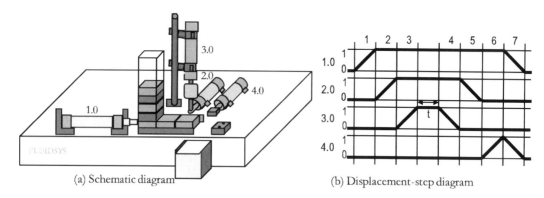

(a) Schematic diagram   (b) Displacement-step diagram

Figure 4.14 | Processing of workpieces.

Problem No. 3 | Pneumatic Flanging Device

A pipe is to be flanged in two steps in a pneumatic flanging device, as shown in Figure 4.15(a). Cylinder 2.0 is in the extended position initially to act as a stop for the advancing pipe. Cylinder 1.0 extends and clamps the pipe, and cylinder 2.0 retracts. Flanging cylinder 3.0 extends for the first time for the preliminary flanging operation and then retracts. The tool switchover is performed by cylinder 4.0, and flanging cylinder 3.0 is in action for a second time for the final flanging operation. The complete retraction of cylinder 3.0 triggers the return motion of clamping cylinder 1.0 and tool switching cylinder 4.0 and the forward motion of position cylinder 2.0. The required sequence of operation is shown in the sequence diagram [Figure 4.15(b)]. Develop a pneumatic control circuit to implement the control task.

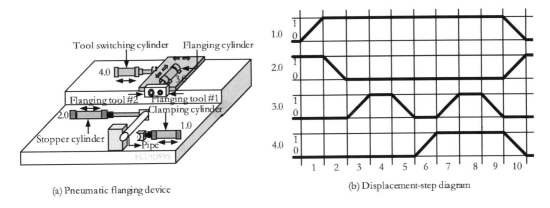

(a) Pneumatic flanging device   (b) Displacement-step diagram

Figure 4.15 | Pneumatic flanging device

Problem No. 4 | A Pneumatic System for Reaming of Drill Holes

A pneumatic system is designed for reaming drill holes, as shown in Figure 4.16(a). Cylinder 1.0 clamps the workpiece. Feed cylinder 2.0 moves fully forward and then backward with a facility for controlling the speed differentially. The feed cycle is repeated two more times. The required sequence of operation is shown in the sequence diagram [Figure 4.16(b)]. Develop a pneumatic control circuit to implement the control task.

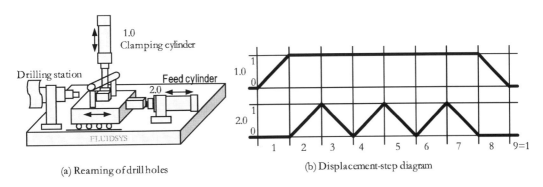

(a) Reaming of drill holes   (b) Displacement-step diagram

Figure 4.16 | Pneumatic system for reaming of drill holes
[Note: Solutions to Additional problems P1 to P4 are given on Page Nos. 60 and 61]

Solutions to Additional Exercises (Given on pages 58 to 59)

A Solution to Problem No. 1

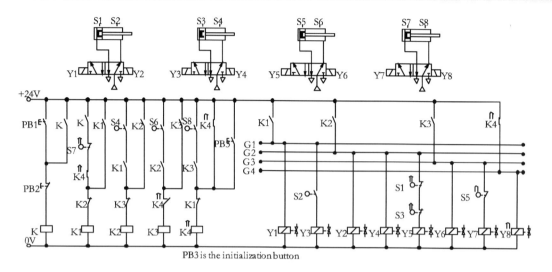

Figure 4.17

A Solution to Problem No. 2

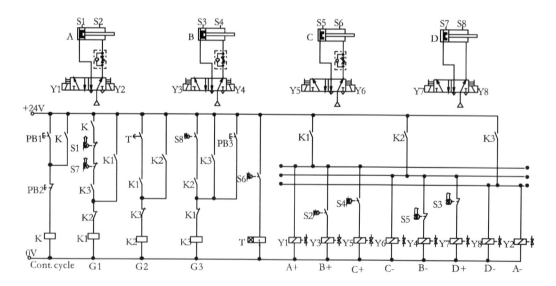

Figure 4.18

A Solution to Problem No. 3

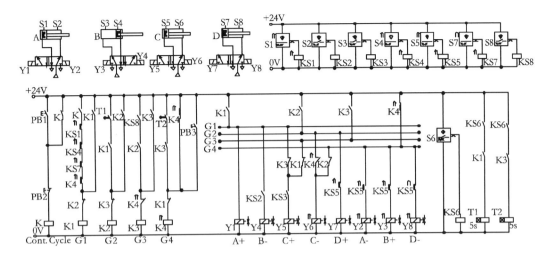

Figure 4.19

A Solution to Problem No. 4

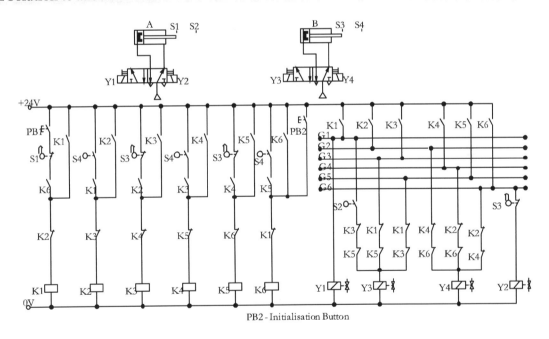

PB2 - Initialisation Button

Figure 4.20

5 | Review Questions

1. What are the advantages of integrating pneumatic and electrical technologies in developing industrial control systems?
2. Briefly explain how a pneumatic valve can be actuated electrically
3. Differentiate between AC solenoids and DC solenoids
4. Give the symbols to ISO 1219
    - 3/2 DC single solenoid valve (NC type), spring return
    - 5/2 DC double solenoid valve with manual override
5. Explain the functional difference between a 'NO contact' and an 'NC contact' in a relay
6. Give the symbols for the following:
    - NO contact, relay
    - NC contact, relay
    - Change-over contact, relay
    - Pushbutton station with 2 NO + 2 NC
7. Explain the working principle of an electromagnetic relay
8. Draw the symbol for an electro-magnetic relay with 3 NO + 1 NC
9. What are the functions of the sensors?
10. List a few applications of sensors
11. Classify sensors giving one example of each
12. What are the disadvantages of contact-type sensors?
13. Explain the working principle of a limit switch
14. Explain the working principle of a reed switch
15. Explain the working principle of the inductive-type proximity sensor giving its block diagram
16. Explain the working principle of a capacitive-type proximity sensor giving its block diagram
17. Explain the working principle of an optical-type proximity sensor giving its block diagram
18. Draw the symbols for the following:
    - Limit switch
    - Inductive-type proximity sensor
    - Capacitive-type proximity sensor
    - Optical-type proximity sensor
19. Differentiate through-beam sensors and diffuse sensors
20. Explain the function of an on-delay timer with a suitable circuit
21. Explain the function of an off-delay timer with a suitable circuit
22. Draw an electrical circuit for the two-hand safety operation with anti-tie down and anti-repeat features. (Hint: Use two pushbuttons, each with two NO contacts, a relay, and an on-delay timer)
23. Explain the function of a pressure switch
24. Explain the function of an up counter
25. Explain the function of a down counter
26. Give symbols for the following:
    - On-delay timer with 1 NO + 1 NC
    - Off-delay timer with 1 set of CO contacts
    - Pressure switch with 1 set of CO contacts
    - Up-counter
    - Down-counter

Appendix 1

Graphic (Symbolic) Representation of Directional Control Valves

Symbols represent pneumatic components because the representations of their complex control functions by sketches are too difficult to draw. The symbols are described in the standard ISO 1219.

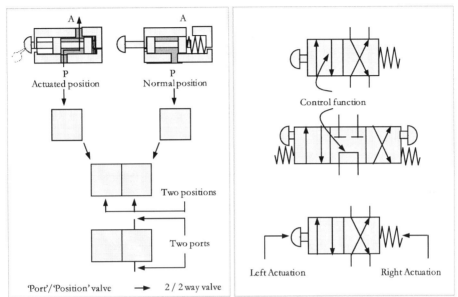

Figure A1.1 | Graphic representation of directional control valves

Figure A1.1 shows how directional control valves can be represented graphically. A pneumatic DC valve is specified as a 'port/position valve', where the 'port' represents the number of ports, and the 'position' represents the number of valve switching positions. Thus, a 3/2-DC valve has three ports and two switching positions. The lines inside the valve represent the function of the valve. The methods of actuation of the valve are shown on the left and right sides of the valve.

Symbols for Basic DC Valves

Symbols aid in the functional identification of the components in the circuit diagrams of fluid power systems. Symbols of basic DC valves are given in Figure A1.2.

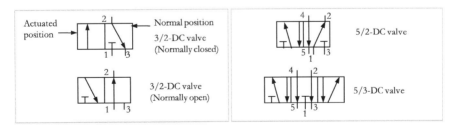

Figure A1.2 | Symbols of basic DC valves

Port Markings

As per the current practice, ports of pneumatic valves are designated using a letter system following the ISO 5599 standard. Figure A1.3 presents the designations for port markings as per the standard. All the inputs and outputs of DC valves must be identified to avoid faulty connections.

Table A1.1 | Port markings

Port	Letter system (Old system)	Number system (As per ISO 5599)	Comment
Pressure port	P	1	Supply port
Working port	A	2	3/2-DC valve
Working ports	A, B	4, 2	4/2- or 5/2- DC valve
Exhaust port	R	3	3/2-DC valve
Exhaust ports	R, S	5, 3	5/2-DC valve
Pilot port	Z or Y	12	Pilot line (flow 1 -> 2)
Pilot port	Z	14	Pilot line (flow 1 -> 4)
Pilot port	Z or Y	10	Pilot line (flow closed)

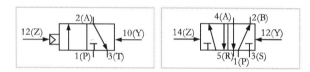

Figure A1.3 | Port markings of DC valves

Methods of Valve Actuation

An essential feature of the directional control valves is their actuation method. These valves can be actuated manually, mechanically, pneumatically, or electrically. Figure A1.4 gives the symbols of various actuating methods of DC valves.

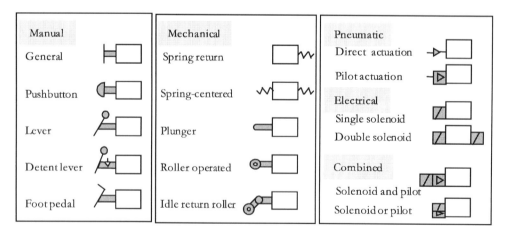

Figure A1.4 | Methods of valve actuation

Appendix 2

Graphic Symbols for Pneumatic Components

A list of the important graphic symbols for pneumatic components is given below.

A2.1 | Supply Elements

Symbol	Description
	Compressed air supply
	Compressor
	Variable compressor
	Air pressure reservoir
	Pressure regulator
	Cooler
	Air dryer
	Air filter
	Manual drain filter
	Automatic drain filter
	Lubricator
	Air service unit
	Air service unit, simplified symbol

A2.2 | Pneumatic Actuators

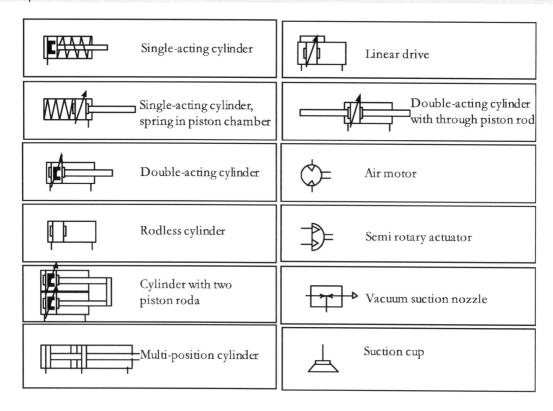

A2.3 | Pneumatic Valves

Symbol	Description	Symbol	Description
	2/2-way valve (NC) type		Check valve
	2/2-way valve (NO) type		Spring-loaded check valve
	3/2-way valve (NC) type		Orifice valve, adjustable type
	3/2-way valve (NO) type		Shuttle valve
	4/2-way valve		Two-pressure valve
	5/2-way valve		Quick-exhaust valve
	5/3-way valve, all closed centre position		Throttle valve with constant restriction
	5/3-way valve, open exhaust centre position		Throttle valve with adjustable flow control
	5/3-way valve, open pressure centre position		Throttle check valve, adjustable

A2.4 | Miscellaneous Components

——	Working line	⊢	Energy tapping point closed
—	Control line	⊘	Pressure gauge
⌒	Flexible line	⊗	Pressure indicator
┴ ┼	Line connection, rigid	⌀	Flow meter
┼ ┤	Line crossover	→∣←	Quick release coupling uncoupled without check valve
▯↧	Exhaust without pipe connection	→∣←	Quick release coupling connected without check valve
▯↧	Exhaust with pipe connection	⊸∣⊷	Quick release coupling uncoupled with check valve
▯	Silencer	⊸⊷	Quick release coupling connected with check valve

A2.5(a) | Electrical Components

Symbol	Description	Symbol	Description
	Normally open contact (NO)		General manual switch contact
	Normally closed contact (NC)		Relay contact
	Change-over conttact		Pushbutton contact
	Switch with NO contact (not automatically reset)		Pull button contact
	Switch with NC contact (not automatically reset)		Twist switch contact
	Mechanically linked contacts		Roller switch contact
	Normally open contacts, actuated		Delay to operate
	Pressure switch		Delay to reset
	Switch with NO contact manually-actuated by rotating		Proximity switch

A2.5(b) | Electrical Components

Symbol	Description	Symbol	Description
—	Direct current (DC)		Relay coil
∼	Alternating current (AC)		Relay coil with delayed contact operation
≂	AC or DC		Relay coil with delayed contact reset
+ −	Positive and negative polarity		Solenoid coil
	Line to earth		Solenoid actuation
	Line to chassis		Solenoid actuation with manual override
	Cell		Double solenoid
or 12 V	Battery		Single solenoid with spring return
24V 0V	Supply and return lines		Solenoid and pilot actuation with manual override
	Fuse		Pnneumatiic – Electric (PE) converter

A2.5(c) | Electrical Components

	Resistor		Light emitting diode
	Potentiometer		Photo-transistor
	Inductor		Opto-coupler
	Inductor with core		NPN transistor
	Variable inductor		PNP transistor
	Capacitor		Triac
	Polarised capacitor		Thyristor
	Variable capacitor		Measuring instruments for current, voltage, resistance and power
	Diode		Motors ac and dc
	Zener diode		transformer

6 | References

1. Joji P., Pneumatic controls, Wiley India Pvt Ltd, New Delhi, 2008
2. FESTO, Electro-pneumatics – Basic Level
3. Textbook 'Pneumatics, Electro-pneumatics – Fundamentals', Frank Ebel, Siegfried Idler, Georg Prede, Dieter Scholz
4. FESTO manuals

Fluid Power Educational Series Books

1. Pneumatic Systems and Circuits -Basic Level (In the SI Units)
2. Industrial Pneumatics -Basic Level (In the English Units)
3. Pneumatic Systems and Circuits -Advanced Level
4. Electro-Pneumatics and Automation
5. Design of Pneumatic Systems (In the SI Units)
6. Design Concepts in Pneumatic Systems (In the English Units)
7. Maintenance, Troubleshooting, and Safety in Pneumatic Systems
8. Industrial Hydraulic Systems and Circuits -Basic Level (In the SI Units)
9. Industrial Hydraulics -Basic Level (In the English Units)
10. Hydraulic Fluids
11. Hydraulic Filters: Construction, Installation Locations, and Specifications
12. Hydraulic Power Packs (In the SI Units)
13. Power Packs in Hydraulic Systems (In the English Units)
14. Hydraulic Cylinders (In the SI Units)
15. Hydraulic Linear Actuators (In the English Units)
16. Hydraulic Motors (In the SI Units)
17. Hydraulic Rotary Actuators (In the English Units)
18. Hydraulic Accumulators and Circuits (In the SI Units)
19. Accumulators in Hydraulic Systems (In the English Units)
20. Hydraulic Pipes, Tubes, and Hoses (In the SI Units)
21. Pipes, Tubes, and Hoses in Hydraulic Systems (In the English Units)
22. Design of Industrial Hydraulic Systems (In the SI Units)
23. Design Concepts in Industrial Hydraulic Systems (In the English Units)
24. Maintenance, Troubleshooting, and Safety in Hydraulic Systems
25. Hydrostatic Transmissions (HSTs) (In the SI Units)
26. Concepts of Hydrostatic Transmissions (In the English Units)
27. Load Sensing Hydraulic Systems (In the SI Units)
28. Concepts of Load Sensing Hydraulic Systems (In the English Units)
29. Electro-hydraulic Proportional Valves
30. Electro-hydraulic Servo Valves
31. Cartridge Valves
32. Electro-hydraulic Systems and Relay Circuits
33. Practical Book: Pneumatics - Basic Level
34. Practical Book: Electro-pneumatics - Basic Level
35. Practical Book: Industrial Hydraulics – Basic Level
36. Programmable Logic Controllers and Programming Concepts
37. Compressed Air Dryers
38. Hydraulic Circuits – Identification of Components and Analysis

For more details, please visit: https://jojibooks.com

Made in the USA
Middletown, DE
10 October 2023

40271810R10046